SKYSCRAPER MANUAL

From concepts to construction methods

COVER IMAGE: **An amazing rooftop view of Dubai Marina skyscrapers, Dubai, United Arab Emirates.** *(Shutterstock)*

First published in June 2020

A catalogue record for this book is available from the British Library.

ISBN 978 1 78521 109 6

Library of Congress control no. 2019909641

Published by Haynes Publishing,
Sparkford, Yeovil, Somerset BA22 7JJ, UK.
Tel: 01963 440635
Int. tel: +44 1963 440635
Website: www.haynes.com

Haynes North America Inc.,
859 Lawrence Drive, Newbury Park,
California 91320, USA.

Printed in China.

Senior Commissioning Editor: Jonathan Falconer
Copy editor: Michelle Tilling
Proof reader: Penny Housden
Indexer: Peter Nicholson
Page design: James Robertson

Acknowledgements

Thanks first to Haynes, particularly Louise and Jonathan, for commissioning the book and for showing compassion, patience and commitment to seeing it through when life's distractions tried to hamper writing efforts.

Credit also to those who make valiant efforts to create publicly available resources that helped in the making of this book. Particularly worthy of mention are the Council on Tall Buildings and Urban Habitat; Designing Buildings Wiki; The Skyscraper Museum of New York; and Wikimedia Commons contributors.

Colleagues and friends Mark Hansford and Anthea Carter provided unfettered support, quite simply with their excellent knowledge but also in offering inspiration and motivation with their infectious enthusiasm. The book's technical illustrator, Anthea, always manages to bring to life often perplexing ideas with her beautifully clear illustrations.

Thanks also to my extraordinary family and friends. My mum, Lis, Si, Oliver and five amazing boys – Tom, Jonty, Alex, Tobias and Ethan – all deserve my love and thanks. But so too do some fantastic friends: Tom, Margo, Jane, Michelle, Diego, Phil, Sarah, Stella, James, Sasha. Thank you for all that you do for me, all of the time.

Finally, I dedicate the book to my beloved Mark, whose enduring kindness, patience and love make the bright times more brilliant and the trickier times so much lighter.

Alexandra Wynne

About the authors

Alexandra Wynne is a journalist, writer and podcast host with a taste for making engineering and architecture accessible and entertaining. She also has over a decade's experience working on international engineering publications.

Dave Parker FICE is a former Visiting Professor of Civil Engineering at Queen's University Belfast and is Technical Editor Emeritus of *New Civil Engineer* magazine.

Anthea Carter is a technical and engineering illustrator. She has freelanced for 30 years since leaving college, working for *New Civil Engineer* and other major engineering publications. www.lunestudio.co.uk

SKYSCRAPER MANUAL

From concepts to construction methods

Builders' Workshop Manual

Dave Parker and Alexandra Wynne

Contents

OPPOSITE **Drone's-eye view of the Hong Kong cityscape.** *(Shutterstock)*

BELOW **The Chrysler Building on the East Side of Midtown Manhattan in New York City.** *(Shutterstock)*

Introduction

Why build tall?

Thousands of years ago, people of the Stone Age dragged huge stones long distances across unforgiving landscapes and set them up in stone circles, or henges. Ancient Egyptians devoted vast resources over many years to the construction of ever larger and higher pyramids. Much later, and across much of the Old World, cathedral spires, minarets and pagodas reached upwards.

For millennia the tallest structure in the world was the Great Pyramid of Giza outside Cairo. Later it was a spire on an unfashionable cathedral in the east of England that took the record (*see* Ancient 'skyscrapers', p.12). What motivated so many people over so many centuries? What drove their obsession with building as tall as the technology of the time would allow?

From the Stone Age to the 19th century, from Europe to the East, the driving force was spirituality. Standing stones, steeples, minarets and pagodas rose towards the sky – the home of the gods, the heaven where the souls of the blessed dead would reside for all eternity. There was also a belief that the taller the stone or spire, the greater the blessing that would be bestowed upon those who contributed to its creation.

These towers were not habitable buildings by any stretch of the imagination. Nor were the Washington Monument or the Eiffel Tower, both subsequent height record holders. It was not until a massive blaze burned out the centre of the booming city of Chicago, creating a prime location for redevelopment that was hemmed in by existing infrastructure, that the first tall buildings of the modern age started to rise (*see* The origins of the modern skyscraper, p.25).

LEFT First attempts to build tall. This Stone Age menhir in northern France stands 9m (30ft) tall. *(China-Crisis)*

For similar reasons tall buildings also took off on the island of Manhattan, the business centre of New York. Restricted space had long driven the construction of taller buildings – such as the 14-storey tenements lining Edinburgh's steep-sided Royal Mile – but in Chicago and Manhattan additional factors came into play.

Both cities were booming, and there was an ever-increasing demand for high-end office space in prestige locations. New technologies, such as cheap structural steel, safety elevators and ways of constructing deep foundations in waterlogged ground, made much taller buildings not only possible but also economically viable (*see* The three key developments, p.19)

Prime building plots in city centres were very expensive. The more office floors that could be stacked up on a given plot, the more the developer could collect in rents. Building tall is also very expensive, however, so a balance had to be struck between height and income. This balance could be hard to achieve.

Unfettered profit-driven development in the

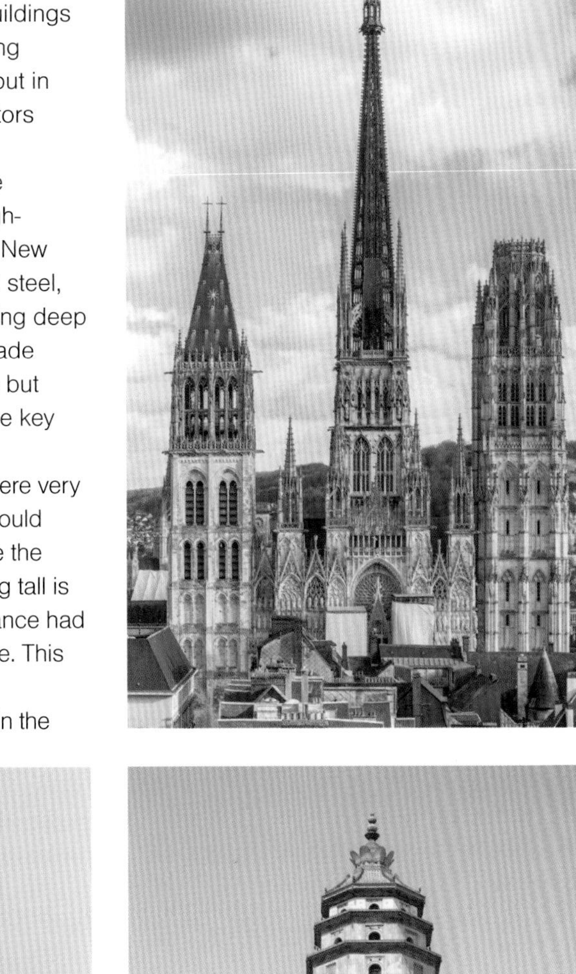

LEFT In France, Rouen Cathedral and its 19th-century 151m (495ft) cast iron spire held the record for the world's tallest building for four years. *(DXR)*

FAR LEFT Measuring 72.5m (238ft) tall, the Qutub Minar in India dates from the 13th century and is the tallest brick-built minaret ever constructed. *(Aazam4u)*

LEFT China's Liaodi Pagoda was built in the 11th century and reaches 84m (276ft). *(Angus Cepka)*

last decades of the 19th century had a negative effect on cities and their inhabitants (*see* The origins of the modern skyscraper, p.26). Regulations that set height limits were introduced in several cities. Nevertheless, the same pressures that triggered the first tall building booms existed. There was also something new developing in the first decades of the 20th century – the urge to build the tallest.

The 'Race to the Sky', as it was dubbed, saw Chicago and New York striving to have the tallest building on the planet, a competition that was finally won by New York's Empire State Building. As a symbol of New York and an iconic building famous throughout the world, it was a great success. As an economic project, though, it was a disaster (*see* Empire State Building, 1931, p.126).

Opened during the Great Depression, it struggled to find tenants and stood half-empty for several decades. It might be expected that the Empire State's problems would deter other developers with equally ambitious plans, but not a bit of it. According to some, there is a common factor in almost all subsequent record-breaking skyscraper projects. They open just at the start of an economic recession, and struggle to find tenants for years afterwards.

Over the decades many skyscrapers have been built in locations where building land is widely available and relatively cheap, where the local demand for office space could easily be met by low- or medium-rise buildings. Yet skyscrapers are still built, even though earlier skyscrapers stand half empty. There is just something powerfully compelling about the concept of building taller than any rivals, about creating an iconic building that will, it is hoped, draw in prestige tenants who believe it will be easier to attract higher-quality staff. Sometimes this works as intended.

Purely commercial development rarely produces landmark skyscrapers. Those that do achieve iconic status and global fame are often the results of rich global

LEFT **Europe's tallest building is currently the 462m (1,516ft) Lakhta Center in St Petersburg.** *(Ad Maskens)*

corporations in search of something special for their new headquarters, a building that will win architectural awards and garner massive media coverage. A world-famous architect and leading structural engineers are commissioned, expensive, high-quality materials such as marble and bronze are specified, budgets are generous.

Prime examples of such projects are New York's Seagram Building and the Gherkin in London (*see* Seagram Building, 1958, p.131, and The Gherkin, 2004, p.145). Admittedly, as time passes, ownership of such buildings can change, and they will have to compete for commercial tenants with other less iconic skyscrapers.

Another motive may explain why almost all the tallest buildings in Europe are located in Russia, with several more in Turkey, and why most record-breaking towers in the 21st century have risen far away from Chicago or London. Some might dismiss many of these supertall skyscrapers as vanity projects, simply part of another 'Race to the Sky'. There seems to be another factor, though.

As new megacities boom – even those without restrictions of space for low-rise developments – city and national governments look to supertall and quirky structures (*see* Super, mega and quirky – the battle to stand out from the crowd, p.61) to symbolise pride in their cities' success and to publicise it to the world.

As such cities boom, they suck in economic migrants by the million. Soon, most of the world's inhabitants will live and work within metropolitan boundaries. Uncontrolled urban sprawl could be the result – as seen on the West Coast of the USA. The alternative – climate catastrophe permitting – is more skyscrapers, that combine residential and mixed use. If these are well designed buildings, part of an integrated urban plan, then skyscrapers are the way to go. The technology exists to make the next generation of skyscrapers much greener, and there is growing public pressure on developers to make sustainability a top priority (*see* Towards the zero-carbon skyscraper, p.169). Life in the sky may become the new normal in the latter half of this century.

DEFINITIONS AND DIFFERENCES

What is a skyscraper?

According to Wikipedia, a building has to be at least 150m (492ft) tall, with more than 40 habitable floors, to be recognised as a skyscraper. Other authorities classify any building below 100m (328ft) as no more than a high-rise building, while anything higher is officially a skyscraper.

As the 21st century unfolds and designers develop ever-taller towers, new categories have been needed. The Council on Tall Buildings and Urban Habitat (CTBUH) broadly recognises a building of 50m (160ft) with 14 storeys as tall. Anything over 300m (984ft) but below 600m (1,969ft) is now classified as a supertall skyscraper. Megatall skyscrapers are buildings over 600m (1,969ft) – although with buildings more than 1km (3,281ft) under construction, another new category might soon be needed.

Structural engineers have a different viewpoint. They consider any building more challenged by wind forces than earthquakes or the loads within it to be a potential skyscraper.

Few tall buildings of the late 19th and early 20th centuries would be classified as skyscrapers by current definitions. Nevertheless, within this manual, the importance of these earlier structures will be celebrated. Elsewhere the 100m (328ft) cut-off point will be generally applied.

Lifts or elevators?

In the UK an elevator is a lift. In the USA and much of the rest of the world a lift is an elevator. In this manual we will use lift when discussing British buildings and elevator elsewhere.

Who's counting?

The first floor in the USA is at street level and immediately above any basement. In the UK this is known as the ground floor. So first floor in a UK building would be the second floor to Americans. For the purpose of this manual we will again use either system as appropriate.

Chapter One

A brief history of the skyscraper

In this chapter we look in detail at the evolution of the skyscraper from the Stone Age onwards. We find out where the term 'skyscraper' came from and where the first tall buildings to be dubbed skyscrapers were built. Also included are the three key inventions that made the skyscraper as we know it possible.

OPPOSITE **One of Bologna's surviving towers leans more than the Tower of Pisa.** *(Maretta Angelini)*

RIGHT The Derby-winning racehorse Skyscraper. *(Public domain)*

Why 'skyscraper'?

In 1789 a racehorse called Skyscraper won the British Derby, the first recorded use of the term. The origin of the name is unclear, but it might be related to the name of Skyscraper's sire, the undefeated stallion Highflyer. Five years later it appears again as the name of a little-used sail on fast square-rigged ships such as tea clippers.

In light winds extra sails were hoisted to the mastheads. Depending on the wind direction, such sails could be square rigged on the masts and were known as moonrakers. Triangular staysails could also be set between the mastheads – these were the skyscrapers.

Subsequently the term was applied to high-flying birds and very tall men! It was not until the late 19th century that the ten-storey Home Insurance Building in Chicago received the skyscraper label. Now, of course, only much taller buildings are classed as skyscrapers.

Ancient 'skyscrapers'

Imperial Roman high-rise

For much of human history people lived in buildings that were rarely more than two storeys high. It was not until Imperial Rome's Golden Age dawned in 27BC that anything taller appeared, in response to a rapidly growing population.

Wealth poured into Imperial Rome from all over the Western world – and beyond. With it came the immigrants, from Italy, from as far away as Britain, seeking wealth, security and all the opportunities a big city could offer. At its peak the population of Rome topped one million – and everyone needed somewhere to live.

Such was the armed might of the Roman legions that the city could abandon its original Servian defensive walls and allow them to decay for centuries. Nevertheless, the city only slowly expanded outwards. Almost everybody preferred to squeeze into the city centre, where Rome's aqueducts delivered clean, fresh water, while a drainage system kept the urban stink down. *Tabernae* selling wine and street food were found on almost every corner. Everyone wanted to live there. In response, private developers began to fill almost any open space with blocks of flats up to nine or even ten storeys high.

Built out of timber, brick or, occasionally Roman concrete, these blocks were known as *insulae* – literally 'islands'. Typically, the ground floor would be occupied by a *taberna*, a shop or a workshop. The more upmarket *insulae* might have running water and sanitation on the lower floors. Most occupants of *insulae*, however, would have to descend to street level to access fresh water and sanitary facilities.

There were no zoning laws in Imperial Rome, so *insulae* could be built in close proximity to the luxurious homes of the rich upper classes. Unscrupulous landlords often insisted on the cheapest of materials and the minimum of maintenance: collapses and disastrous fires were commonplace.

This meant that, unlike modern high-rise flats, the better-off occupied the lower floors where there was a better chance of escape in the event of a disaster. Those who could only afford the cheapest rents had to climb many flights of stairs to the invariably primitive upper storeys.

To reduce the risks of more *insulae* catastrophes, laws were passed limiting their height to 20.7m (67.9ft) initially, and then 17.75m (58.2ft) after the Great Fire of Rome in AD64. These restrictions seem to have been widely ignored. It is estimated that the number of *insulae* in the city peaked at around 48,000, and there were more *insulae* in other major Roman cities.

By AD270 Rome's thinly stretched legions could no longer be relied upon to crush any barbarian invaders well away from the city. A new, much larger and stronger wall was built, which survives to this day.

A 'Desert Manhattan'

City walls may offer security, but they are also something of a developmental straitjacket. In the deserts of Yemen the 160km (99 mile)-long Wadi Hadramawt was in medieval times a ribbon of green up to 2km (1.2 miles) wide, kept fertile by twice-yearly monsoon floods

LEFT Roman *insulae* could be as tall as ten storeys high. Collapses were common. *(Public domain)*

and complex irrigation systems. Known since Roman times for its luxuriant groves of frankincense and myrrh trees, the wadi was also a major trade route between the Red Sea and the Mediterranean. Towns and cities sprang up along its length. All needed defensive walls, as inter-tribal warfare was endemic and the desert nomads were a constant threat to farmers and merchants alike.

One such was the walled city of Shibam. For centuries it served as the commercial centre of the wadi. Frequently attacked, conquered at least twice and razed to the ground in the early 10th century, it was also devastated by flash flooding in the 13th and 16th centuries. Eventually it was rebuilt on a low mound above the wadi, surrounded by a defensive wall that enclosed as little of the precious fertile land as possible.

As the city flourished, the threat of attack still remained. So, within the confines of the wall, towers began to rise, some as high as 29m (95ft), with as many as 11 floors. Construction was almost entirely of mud bricks. Few large trees grew in the region; timber suitable for building was scarce and expensive. Readily available timber came in short lengths and had poor durability. At best the central stair core might have rubble stone lower levels.

Typically the first two storeys would be windowless, with the ground floor used for food storage and large animal stalls and the floor above housing smaller animals such as goats, sheep and rabbits, as well as more storage. The public reception rooms were on the second floor. Above this were the private family rooms, and above them the roof was usually set back to create screened terraces.

Narrow alleyways ran between the towers, so narrow that bridges could connect the upper floors of adjoining towers, allowing women to visit each other without the inconvenience of descending to ground level and donning the veil.

Mud-brick construction is not noted for

BELOW Shibam – Yemen's 'Desert Manhattan' – is virtually unchanged since the 16th century. *(Jialiang Gao)*

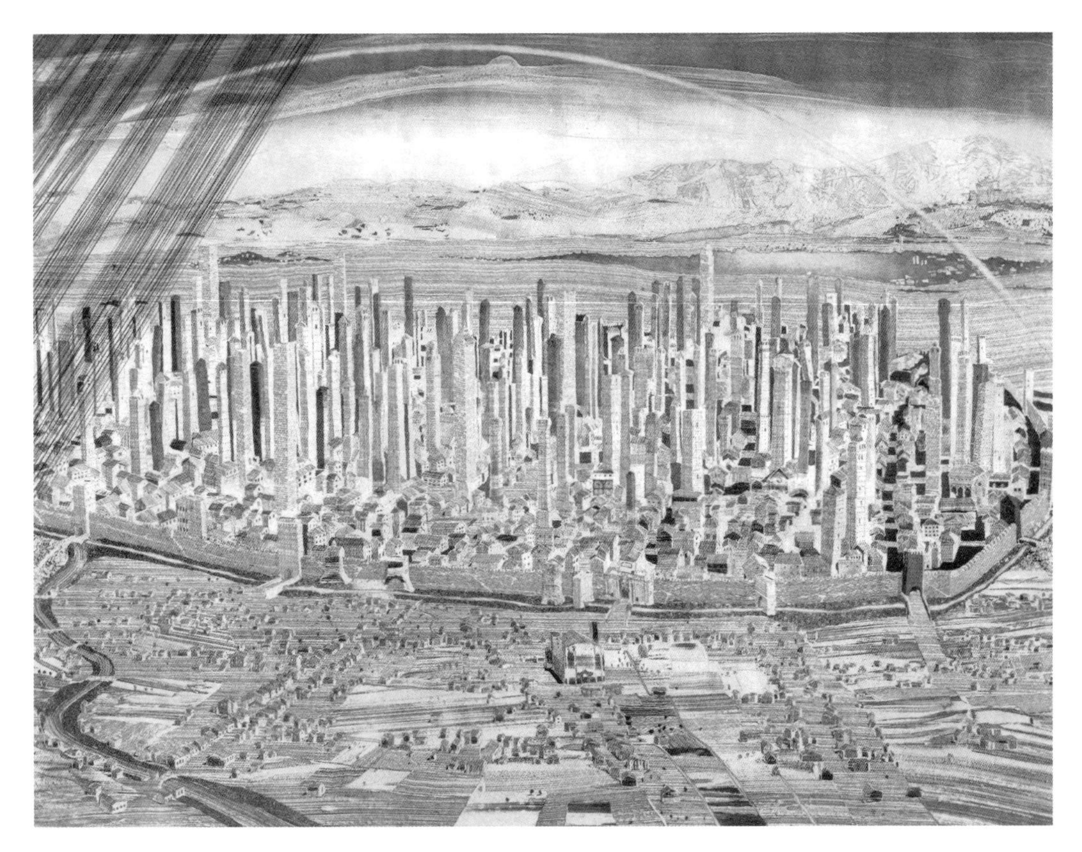

LEFT No one is really sure what medieval Bologna's many towers were used for. *(Toni Pecararo)*

its durability. Over the centuries most of the original towers crumbled and had to be replaced. Strict local ordinances meant that the replacements had always to conform to the original pattern. Shibam today looks much as it did four or five centuries ago, with around 500 towers and the highest crowded around the edge of the mound immediately behind the encircling wall. This is now reduced to little more than waist height.

One of the first non-Arabs to discover Shibam was the famous explorer, Dame Freya Stark, during one of her expeditions to the desert in the 1930s. It was she who dubbed the city 'The Manhattan of the Desert', an appellation that has stuck. At the time of publication (2020), however, the conflict in Yemen remains a threat to the ancient city. Mud bricks offer little resistance to modern weapons.

Bologna's mysterious towers

There is at least one other city that could claim to be a 'medieval Manhattan' – Bologna in northern Italy. Between the 12th and 13th centuries a forest of slender towers sprang up within the city walls – and nobody really knows quite why.

At any time during this particular period of history there were 100 or more such towers crowded together. Some were nearly 100m (328ft) tall. Their function remains obscure. Some have suggested they were lookout towers, built at a time of chaos and conflict – but why so many? Others speculate they had some defensive function, perhaps as refuges in case invaders breached the city walls.

Against that viewpoint, however, is the fact that the surviving towers show few signs of the defensive features common on medieval strongholds, such as arrow slits and battlements. Most are so slender that they would be extremely inefficient and inconvenient habitations – not to mention the lack of windows or even internal floors.

It seems that these towers were not built by the city but by individual rich families. During the 12th century these families split into two hostile factions: one supporting the Pope, the other the Holy Roman Emperor. This was at the time of the so-called Investiture Controversy, when the Roman Catholic Church and the Emperor clashed over who had the right to appoint bishops and abbots.

One theory is that the real purpose of the

ABOVE Damaged by an earthquake centuries ago, the Great Pyramid of Giza remained the tallest building on Earth for millennia. *(Nina)*

BELOW Only this model of Lincoln Cathedral's record-breaking spire still remains. Unusually, the spire was a timber-framed, lead-clad structure. *(Aidan McRae Thomson)*

RIGHT The Eiffel Tower is the tallest wrought-iron structure ever built. *(Benh Lieu Song)*

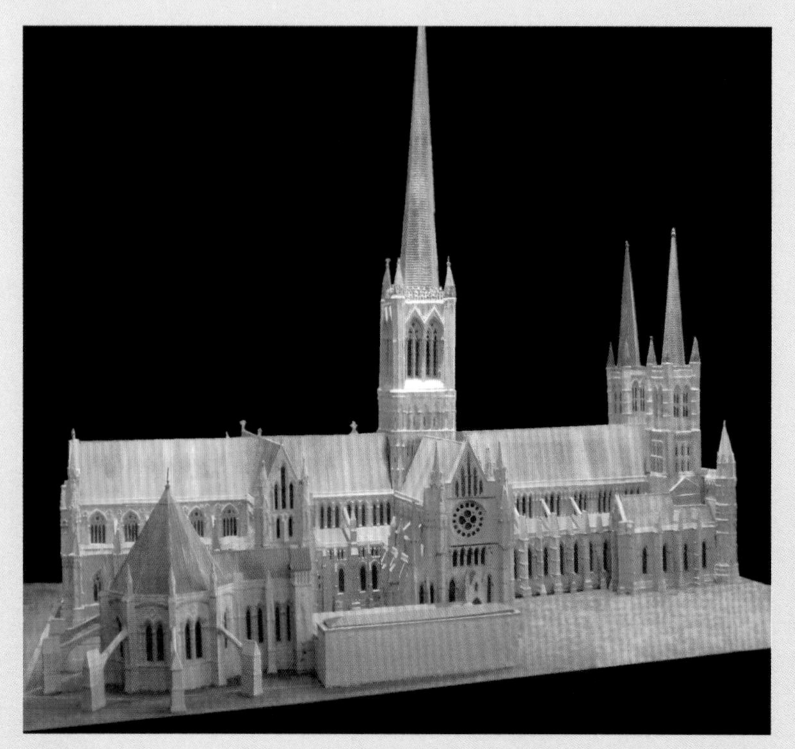

For nearly 4,000 years the tallest man-made structure on the planet was Egypt's Great Pyramid of Giza. With more than 2.3 million limestone blocks forming the core and a smooth white limestone sheath, the pyramid, as built, reached 146.5m (481ft) high. In 1303, however, a major earthquake shook off the sheathing and reduced the height to just 138.8m (455ft). Much of the shattered sheathing was used to construct mosques in nearby Cairo.

A very different structure took the record from the Great Pyramid. Lincoln Cathedral in eastern England had also been struck by an earthquake, this one in 1185, which destroyed virtually all of the original building. Reconstruction was slow: it was not until 1311 that a wood-framed, lead-sheathed spire was added to the central tower. This reached 160m (525ft) and held the record as the world's

tallest structure until it collapsed during a violent storm in 1548. It was never replaced.

Several more cathedral spires claimed the record in turn during the following centuries, but none topped Lincoln's. It was not until 1884 that the Washington Monument, a hollow stone obelisk commemorating the first US President, took the record at 169m (555ft). This reigned supreme for just five years, until Paris's Eiffel Tower smashed all previous records at 312m (1,024ft).

Both these structures are monuments rather than functional buildings. There was only one 19th-century structure that could lay claim to be the world's first skyscraper by current definition – the 167.5m (550ft) Mole Antonelliana in Turin, Italy. Completed in 1889 and originally intended to be a synagogue, the tower is the tallest building ever constructed from unreinforced brickwork and is now home to 'the world's tallest museum'.

LEFT Dubbed 'the world's tallest museum', Turin's Mole Antonelliana is also the world's tallest unreinforced brick building. *(Livioandronicus2013)*

BELOW The Washington Monument tops all these historic structures in this 1884 montage. *(George F. Cram)*

DIAGRAM OF THE
PRINCIPAL HIGH BUILDINGS
OF THE OLD WORLD.

Feet High.

1. Porcelain Tower, Nankin, China ... 200
2. St. George's Hall, Liverpool ... 85
3. Tomb of Theodoric, Ravenna ... 50
4. Chichester Cathedral ... 271
5. Victoria Tower, Westminster ... 331
6. Boston Church, Lincolnshire ... 292
7. Taj Mahal, Agra ... 220
8. York Cathedral ... 198
9. Temple of Bacchus, Teos ... 50
10. Alexandrian Column, St. Petersb'g 154
11. Column of July, Paris ... 154
12. Torre Asinelli, Bologna ... 370
13. Bell Tower, St. Mark's, Venice ... 323
14. Coloseum, Rome ... 157
15. Friburg Cathedral ... 385
16. Temple of the Sun, Baalbec ... 120
17. Temple on the Ilissus, Athens ... 25
18. Erechtheium, Athens ... 35
19. Chartres Cathedral ... 403
20. Church of Ste Genevieve, Paris ... 274
21. The Monument, London ... 202
22. Amiens Cathedral ... 383
23. Church of St. Theobald, Tharin ... 330
24. Royal Albert Hall, London ... 154
25. St. Stephen's Cathedral, Vienna ... 441
26. Torazzo of Cremona ... 396
27. Hotel des Invalides, Paris ... 310
28. Temple of the Giants, Agrigentum 116
29. Parthenon, Athens ... 66
30. Second Pyramid, Gheezeh ... 447
31. Rouen Cathedral ... 460
32. Strasburg Cathedral ... 468
33. Eleanor Cross, Waltham ... 50
34. Cologne Cathedral ... 510
35. Great Pyramid ... 460
36. St. Peter's, Rome ... 448
37. St. Paul's, London ... 360
38. Albert Memorial ... 180
39. { Obelisk, Luxor ... 75 / Prophylon ... 79 }
40. Bow Church, London ... 235
41. Cleopatra's Needle ... 68
42. Old St. Paul's, London ... 508
43. Church of St. Mary, Lubeck ... 400
44. Abbey of St. Stephen, Caen ... 400
45. Church of St. Martin, Landshut ... 460
46. The Baptistry, Pisa ... 190
47. Tomb at Mylasa, Caria ... 50
48. Church of St. Peter, Hamburg ... 380
49. Obelisk in Piazza di San Giovanna in Laterano, Rome ... 153
50. Antwerp Cathedral ... 403
51. "Bell Harry" Tower, Canterbury ... 235
52. Tower of the Winds, Athens ... 45
53. The Cathedral, Florence ... 376
54. Hotel de Ville, Brussels ... 374
55. Mosque of St. Sophia, Constantinople ... 182
56. Pantheon, Rome ... 143
57. Chapel of St. Pietro Montorio, Rome 40
58. Choragic Monument of Lysicrates, Athens ... 34
59. Salisbury Cathedral ... 404
60. Trajan Column, Rome ... 134
61. Cathedral, Frankfort-on-Main ... 326
62. Pyramid of Mycirinus ... 218
63. Church of St Nicholas, Newcastle-on-Tyne ... 201
64. Temple of Jupiter Stator, Rome ... 98
65. Mechlin Cathedral ... 319
66. Bell Tower, Florence ... 266
67. Tomb of Absolom, Jerusalem ... 54
68. Norwich Cathedral ... 309
69. Leaning Tower, Pisa ... 188
70. Pompey's Pillar, Alexandria ... 100
71. Church of St. Isaac, St. Petersburg 336
72. { Central Spire, Lichfield ... 252 / Western Spire ... 192 }
73. Arch of Constantine, Rome ... 70
74. Tower of Ivan Veliki, Moscow ... 260
75. Central Transept, Crystal Palace ... 198
76. Science Schools, South Kensington 110
77. Temple of Vesta, Tivoli ... 55
78. Washington Monument ... 555

The Red Tint indicates BRICK; *the Stone Color,* STONE; *the Pink,* GRANITE; *the Purple,* BRONZE, COPPER *or* LEAD; *and the Yellow,* GOLD.

tower-building boom was to impress rival families with a soaring symbol of a family's wealth and power. Building a tower was not to be undertaken lightly. A tower 60m (197ft) high could take as long as ten years to complete. Most seem to have followed a similar pattern.

In cross-section the towers were square and sat on foundations up to 10m (32.8ft) deep. Above a solid stone base the walls were built in a 'sacco' masonry – an outer and inner skin of brick or stone with the central gap filled with a mix of rubble and lime mortar. This was a universal construction technique in medieval Europe.

Over the centuries most of the original towers collapsed or were demolished. A few of the more squat towers were converted into prisons or residential buildings. Fewer than 20 now remain.

Of these, two have a particular claim to fame. Both tilt away from the vertical, like the much better-known Leaning Tower of Pisa. At 97.2m (319ft) tall the Asinelli Tower is much higher than the 56m (184ft) campanile in Pisa, but only leans a third as much – about 1° away from the vertical. The 48m (158ft) Garisenda Tower, however, leans at 4°, slightly more than the much-photographed Pisa landmark.

In all three cases the tilt is due to inadequate foundations, and many of the recorded collapses during past centuries were down to the same causes.

BELOW Edinburgh's Royal Mile has linked the city's castle with Holyrood Palace since the 12th century. *(Public domain)*

Edinburgh's record-breaking tenements

In the 17th century the tallest inhabited buildings on the planet were probably the tenements in Edinburgh's Old Town. Some reached 14 storeys high, forced upwards by the restrictions imposed mainly by the Old Town's acute shortage of straightforward building land.

Scotland's capital grew up on the steep-sided glacial 'tail' running eastward from the volcanic crag on which Edinburgh Castle stands. What is now known as the Royal Mile occupies the crest of the ridge, sloping down towards the Holyrood Palace, the official residence of the Kings of Scotland. On either side of this road there is very little flat land. By the 16th century two defensive walls had been built, but these enclosed just 57ha (140 acres).

Originally, Edinburgh's population, probably no greater than a few thousand in the 15th century, mostly huddled together in single-storey dwellings clinging to the slopes either side of the Royal Mile. This had risen to 12,000 by the middle of the next century, despite the frequent wars and outbreaks of plague that were so common at the time. It continued to grow. A century later the only way to expand within the walls was upwards.

Tenements differ from modern tower blocks in that they were built on to each other, forming long lines of tall habitations separated at intervals by narrow 'closes' or 'wynds'. Builders often dug deep cellars, sometimes used as extra accommodation or even shops and taverns. As in Imperial Rome, the wealthy and better-off lived in the lower floors. Sanitation in tenements was even more primitive: the dreaded cry of 'Gardy-loo!' from an upper-floor window warned passers-by

LEFT Tenements up to 14 storeys high dominated the Royal Mile until the late 19th century. *(Samuel Dukinfield Swarbreck)*

that a shower of human waste was about to descend upon them.

The city became notorious for its overcrowding and obnoxious stench. With nowhere left within the old boundaries for significant expansion, the city fathers launched ambitious plans to build an elegant, modern and spacious New Town across the valley to the north. The wealthy and professional classes soon abandoned the noisome Old Town, which was left to decay.

Fires and collapses began to nibble away at the ageing tenements and many hapless residents were killed as a result. Then in 1824 came the so-called 'Great Fire of Edinburgh'. Thirteen lives were lost and 400 homes destroyed. Only 24 tenements were totally destroyed, but these included two which were 11 storeys high, and two more which were some of the earliest timber-framed tenements ever built.

New tenements arose, stone-built, rarely more than five storeys tall. Unlike the first-generation tenements these featured large rooms, high ceilings and extensive ornamentation. In the 21st century these are now seen as highly desirable properties, and their future seems secure.

The three key developments

Going up

While the impressive gains of combining steel and masonry into one gave rise to building tall, it is a perhaps less obvious, but no less impressive, mechanism that made it truly desirable to go sky high.

A significant engineering breakthrough by an American inventor named Elisha Graves Otis in the 19th century paved the way for today's superfast elevators that transport people up and around the tallest buildings in the world.

Although elevators were in use in Roman times, it was the arrival of steam power that saw them begin to appear in factories, mines and warehouses. But the system's uptake as a primary means of moving people to great heights was hampered by a poor safety record. At that time if the natural fibre ropes that held the elevator broke, it would simply, and catastrophically, fall.

Enter Otis, who in 1852 developed an ingenious mechanism that would make these deadly devices safer. But trying to convince a wary public of the genius of his idea was not

RIGHT One of the men who made tall buildings possible – Elisha Graves Otis. *(Public domain)*

easy. And so he came up with a way to grab people's attention.

Otis prepared a death-defying stunt at the 1853/54 World's Fair in New York in the hope it would persuade the world at large that his new invention would change how people use buildings forever.

Gathering crowds were treated to quite a display. Legend has it that, with the aid of circus icon Phineas Taylor Barnum, Otis mounted an elevator platform in the centre of New York's Crystal Palace hall and had himself hoisted aloft. Above him an axeman stood poised. Otis gave the order, the axeman swung his long-handled axe and the natural fibre rope supporting the platform was chopped clean through.

There were gasps and cries from the crowd, but the plummeting elevator quickly shuddered

RIGHT Otis revealed his safety brake invention in a death-defying stunt to gathering crowds at the 1853/54 World's Fair in New York. *(Public domain)*

to a halt as the patented safety brake kicked in. Incidentally, although eyewitness accounts seem to agree that a man with a felling axe severed the rope with one mighty stroke there is no photographic evidence one way or another. The accompanying sketch suggests that the rope was sawn through, less dramatically, by a man with a knife.

Otis's invention combined two vertical guide rails fitted with metal teeth with a spring-loaded mechanism on the top of the car. In normal use this mechanism was kept closed by the tension in the supporting cable. Should the cable ever snap, the springs would be released, propelling strong metal hooks outwards and clamping them on to the guide rails.

Otis's game-changing engineering and his attention-grabbing performance saw his safety elevators go into widespread production a few years later.

But the elevators in today's skyscrapers have only a small amount in common with Otis's trailblazing device. Now their safety is based on vastly more advanced designs, developed and refined over decades by highly technical engineers, knowledgeable in not just physics but also in technological innovation.

Modern elevators have multiple cables and friction brakes that resemble those found on cars, alongside a development of Otis's original safety device that latches on to the hoisting rails when needed. Buffers that can help dissipate the energy of a falling elevator and slow it down also back these up.

Systems check and monitor all aspects of the elevator, including whether there is a power cut, if the doors are fully closed and its speed. If anything is not right, the fail-safe kicks in and the safety brakes take hold until the issue is fixed.

Otis's invention fuelled a burning desire to build higher – much higher. With the help of engineers who followed in his footsteps, the true potential of skyscrapers began to be unleashed, enabling cities to expand upwards as well as outwards.

Previously undesirable higher floors have since become prestigious dwellings for workers and residents – paving the way for today's penthouses to fetch millions in the world's most sought-after cities.

ABOVE Modern elevators have multiple cables and friction brakes and many fail-safes built in. *(Shutterstock)*

Digging deep

New York and Chicago are well known for spawning the world's earliest skyscrapers, but forging the development of building tall was not the easiest of undertakings for either city. The proximity of water – from the Hudson River to the former and Chicago River and Lake Michigan to the latter – meant that finding stable ground was a challenge from the outset.

While broad but shallow foundations, similar to those found on lower-rise buildings, had been used for some of the earliest skyscrapers, a number of taller buildings had begun to experience problems with uneven settlement (*see* The origins of the modern skyscraper, p.26).

All buildings sink to some degree as the natural ground gradually settles under the load of the structure above. Engineers manage how much that affects the structure and, heeding this, they design the foundations to suit the height of the building and the kind of ground it stands on. The taller the building, the greater attention is needed to how it is anchored below ground.

The earliest skyscraper builders were faced with the problem of digging deep to find stable strata strong enough to support the building's unprecedented weight. Unfortunately, in both Chicago and Manhattan, these desirable strata usually lay well below the water table,

1 **Steel caisson** is sunk into the ground, pressurised air is fed to the working chamber to prevent groundwater entering

2 **Muck tube** is filled with water, to the same height as the water table, to create a seal

3 **Clamshell bucket** is suspended from a steam crane at the surface and can easily be lowered through this water and retrieve the 'muck' that is being excavated in the working chamber

4 **Air lock** enables workers and equipment to enter the pressurised working chamber

5 **The caisson** is sunk deeper, as it is excavated, by its own weight and extra ballast, then a brick lining is built on top. This whole structure is eventually backfilled with concrete and brickwork to create the foundation

6 **Pneumatic caisson** works on the same principal as a cup holding air, pressed into water upside down, if no air escapes from the cup, little water will enter the cup's interior as air pressure keeps it out

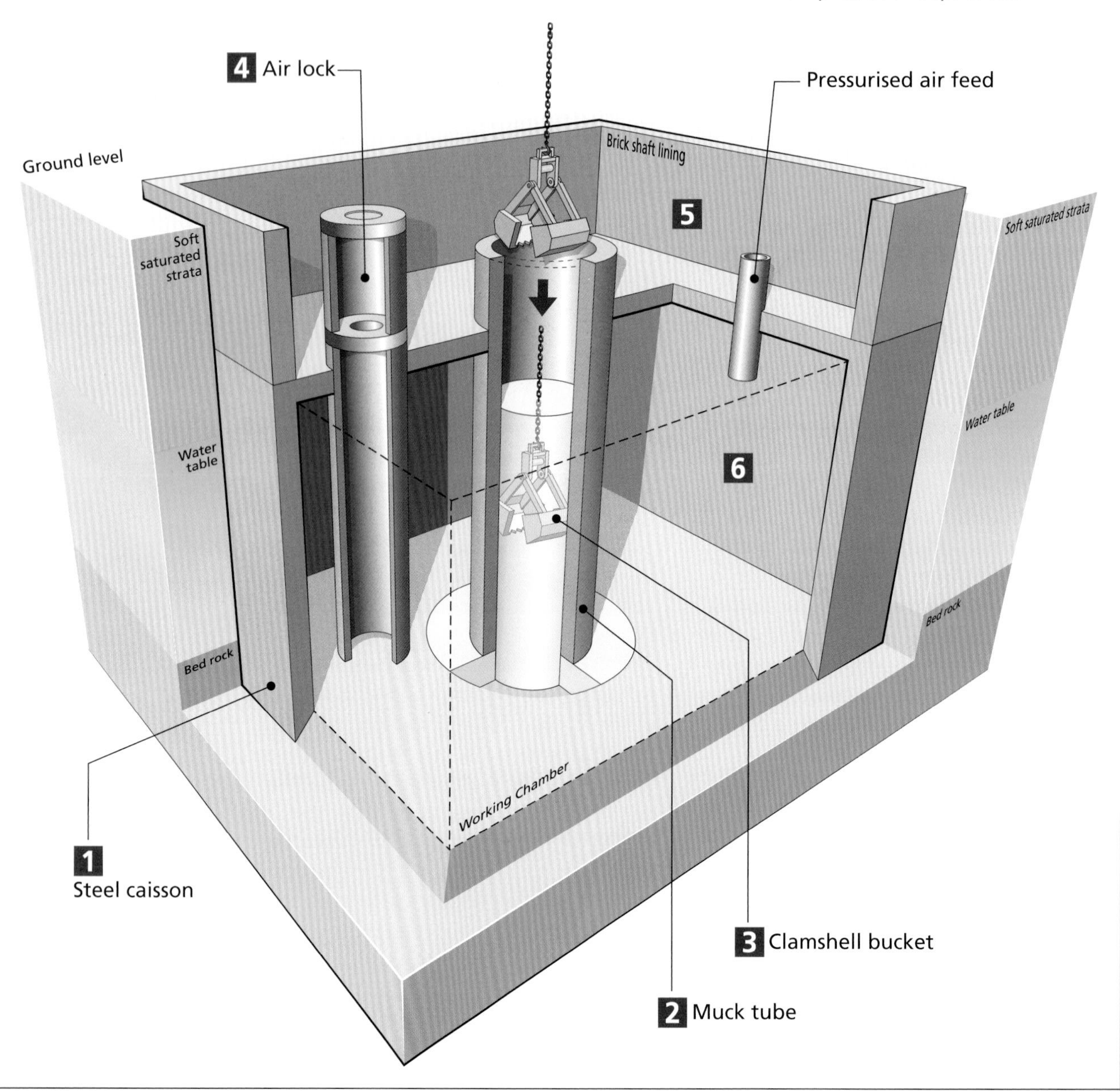

which meant the ground above was weak and saturated.

An intriguing method of constructing foundations in rivers had begun to be developed in the UK for bridge projects – the pneumatic caisson. The term caisson comes from the French word *caisse*, which means box, and a pneumatic caisson is a four-sided or circular box with a roof but no bottom (*see* diagram, p.22). The roof would have one or more holes where shafts enabled workers or materials to pass, via an airlock or water seal, into and out of the working chamber.

Inside, air pressure was kept just high enough to balance the ground water pressure – keeping the water and mud outside while allowing workers to dig out the soft material and transfer it to a central pit. The muck tube would be filled with water up to water-table level – the weight of this water would balance the air pressure in the working chamber, preventing any leaks. A clamshell bucket suspended from a steam crane at the surface could easily pass through this water seal to lift this 'muck' to the surface.

As the workers removed material from under the caisson's perimeter, it would slowly descend. The shaft it created would be lined as it went down, usually with robust brickwork supported by the caisson itself.

Once all the soft material had been removed, the caisson and the shaft would be filled with concrete and brickwork to create solid foundations for a skyscraper.

The 106m (348ft) Manhattan Life Insurance Building in New York, completed in 1894, was the first skyscraper to be entirely founded on pneumatic caissons. Both circular and rectangular caissons were used, fabricated from steel plate and standing 3m (9.9ft) tall. They had to penetrate around 16.5m (54ft) of mud and quicksand to reach bedrock while supporting a masonry shaft lining. Average progress during this stage was 1.2m (3.9ft) per day. Once the bedrock had been smoothed and levelled, the caissons were filled with concrete and topped with solid masonry columns.

But pneumatic caissons have been mired in controversy from the very early days of their use – perhaps most notoriously on the Brooklyn Bridge, built in the 1870s, when the ill effects of working under high air pressure in the inhospitable lower chamber had begun to be more fully understood.

Dubbed 'caisson disease' by Dr Andrew Smith, the physician overseeing the bridge's construction, it is a condition known more familiarly today as the Bends (decompression sickness). Whichever the term, the condition arises when the excess nitrogen that dissolves into the bloodstream under high air pressure bubbles out after a person returns to normal atmospheric pressure. These bubbles can cause intense joint pain, cardiovascular problems or even

OPPOSITE How the early pneumatic caisson worked. *(Anthea Carter)*

BELOW The 106m (348ft)-tall Manhattan Life Insurance Building in New York, completed in 1894, was the first skyscraper founded entirely on pneumatic caissons. *(Public domain)*

THE MANHATTAN LIFE-INSURANCE COMPANY OF NEW YORK.
64 AND 66 BROADWAY (NOW BUILDING), BETWEEN WALL STREET AND EXCHANGE PLACE.

ABOVE The ill effects of working in high air pressure conditions began to be well understood during construction of the Brooklyn Bridge. *(Shutterstock)*

RIGHT Henry Bessemer made low-cost, high-quality steel widely available – paving the way for skyscrapers, railways and ocean liners. *(Public domain)*

death. Many died or were crippled for life before the cause was identified.

The growing awareness of the risk of illness was aided by Dr Alphonse Jaminet, the physician on the Eads Bridge project over the Mississippi River, which had started construction just before the Brooklyn Bridge and used similar caissons. Together Smith and Jaminet began some of the earliest health and safety investigations. They created a number of countermeasures to limit workers' exposure to these adverse conditions, and ultimately reduce the number killed or crippled as more and more adventurous projects needing complicated foundations were pursued.

The man who made steel frames possible

Until the late 19th century the only structural materials available to builders that had not

LEFT An early Bessemer converter in action. *(Public domain)*

been around for millennia were cast and wrought iron.

Cast iron was brittle – it could be used for columns, but not for beams. Wrought iron, an excellent structural material in many ways – the Eiffel Tower is built from it – contains unpredictable quantities of slag, is slow and expensive to produce and its structural properties are variable. Steel had been made for centuries, but in very small quantities, and was used mainly for weapons and tools. (Reinforced concrete was not an option: the first large-scale manufacture of Portland cement in the USA only began in 1886.)

All this began to change in 1856, when a wealthy English inventor by the name of Henry Bessemer was granted his first patent for what became known as the Bessemer Process.

Bessemer had become rich after inventing a 'gold' pigment made from brass, but his main interests lay in the field of weapons technology. What he was searching for was a means of producing high-strength steel cheaply and in large quantities, to mass-produce larger, more powerful artillery. Bessemer's solution was a development of an ancient Chinese technique in which air is forced through molten pig iron to purify it and burn off most of the carbon.

Pig iron is what is produced by a blast furnace. A mixture of iron and slag, it also contains 4% carbon or more. With that much carbon the pig iron is very brittle and requires further processing to transform it into a useful material.

Ultimately it is the carbon content that determines the basic properties of iron-based materials. To make steel, this carbon content has to be reduced to less than 1%. Bessemer's new converters could do this and produce a 'blow' of up to 30t (33 US tons) of low-carbon molten steel in just 20 minutes.

LEFT Immediately after the Great Fire of 1871, Chicago's business centre was largely rebuilt with four- to five-storey traditional masonry buildings. *(Public domain)*

LEFT Although sometimes known as 'the first true skyscraper', the ten-storey Home Insurance Building, built in 1884 and extended by two storeys in 1890, was more of a transitional design. *(Public domain)*

At first, however, he struggled to achieve a reliable product. The only way of estimating how much carbon and slag was left in the blow was by the colour of the flames coming out of the top of the converter. Then came the breakthrough. The blow was continued until all the carbon was burned off, and then the correct amount of carbon plus other strength-enhancing minerals were added.

Dubbed mild steel, this new alloy sent the Industrial Revolution into overdrive. It was strong, consistent and easy to hot roll into steel plates, bars and the structural I-beams that made skyscrapers possible. Steel prices tumbled by 80% and production soared. Railways (railroads) spread across the continents; giant steamships criss-crossed the oceans. And in Chicago and Manhattan building designers began to transform the cityscape. The vertical city was born.

The origins of the modern skyscraper

Rising from the ashes

In October 1871 fire ripped through the centre of Chicago. It was a disaster that was widely predicted. Homes, schools, offices, civic buildings – all were largely built of timber. Chicago had been suffering from a major drought that summer, so the timber in the buildings was tinder-dry. The local press warned of the dangers, but nothing was done.

According to local legend the fire began on a Sunday evening when a cow being milked by a Mrs O'Leary kicked over a kerosene lamp and set fire to the hay in the barn. Fanned by strong winds the fire spread rapidly. Soon 30m (98ft)-high walls of flame were raging through the city centre, consuming everything in their path. By the time long-overdue rain finally extinguished the last of the flames early on Tuesday morning, the centre of Chicago was a blackened scar measuring around 6.4km by 1.6km (4 miles by 1 mile). Around 300 citizens lost their lives, more than 17,000 buildings were destroyed and more than 100,000 people were left homeless.

Chicago had no defensive wall like medieval cities, yet the area for redevelopment was constrained. Lake Michigan and the Chicago River, and the railroads and stockyards on which the city's prosperity was based, and which survived the 1871 blaze, surrounded the burnt-out city centre. In the middle of the 19th century Chicago had become the world's fastest-growing city: after the Great Fire the boom continued. The conditions were right for designers to look skywards, and consider building much higher than the low-rise structures that had been lost.

Inevitably, there was a demand that any new buildings must be fireproof, so timber construction was abandoned.

However, traditional fireproof construction materials such as stone and brick have their limitations. Tall load-bearing masonry walls have to be very thick at their base –

TALLEST SKYSCRAPERS – TIMELINE

The world's tallest skyscraper record holders

1885–1890	**Home Insurance Building, Chicago, 55m (180ft)** ***demolished 1931***
1890–1894	**World Building, New York, 94.2m (309ft)** ***demolished 1955***
1894–1899	**Manhattan Life Insurance Building, New York, 106m (348ft)** ***demolished 1964***
1899–1908	**Park Row Building, New York, 119.2m (391ft)**
1908–1909	**Singer Building, New York, 186.6m (612ft)** ***demolished 1968***
1909–1913	**Metropolitan Life Tower, New York, 213.4m (700ft)**
1913–1930	**Woolworth Building, New York, 241.4m (792ft)**
1930–1930	**Bank of Manhattan, New York, 282.6m (927ft)**
1930–1931	**Chrysler Building, New York, 318.9m (1,046ft)**
1931–1972	**Empire State Building, New York, 381m (1,250ft)**
1972–1974	**One World Trade Center, New York, 417m (1,368ft)** ***demolished 2001***
1974–1998	**Sears (now Willis) Tower, New York, 442.1m (1,451ft)**
1998–2004	**Petronas Towers, Kuala Lumpur, 451.9m (1,483ft)**
2004–2010	**Taipei 101, Taipei, 508m (1,667ft)**
2010-present	**Burj Khalifa, Dubai, 828m (2,717ft)**

and the taller the wall, the thicker the base. An extreme example is the 169m (555ft) granite Washington Monument. Despite the high strength of granite, at their base the monument's walls are nearly 5m (16ft) thick.

As early as the 1860s in England and France there were experiments with internal supporting frames, usually of cast and/or wrought iron, that could stabilise thinner tall masonry walls and take some of the internal floor loads. In 1864 the five-storey Oriel Chambers in Liverpool, England, pioneered the use of glass curtain walling hung off iron columns, which also supported floor loads.

This was the work of local architect Peter Ellis, who also designed a second, similar building in Liverpool. Studying in the same city at the time was a teenage refugee from the US Civil War, a certain John Wellborn Root. On his return to the USA he became an apprentice architect in New York, eventually moving to Chicago in 1871 to work as a draughtsman in an architectural practice busy with post-Great Fire projects. Root eventually became famous as one of the founders of the Chicago School style of tall building design.

One of the challenges designers had to face were the ground conditions in central Chicago. To create land safely above the levels of the Chicago River and Lake Michigan, nearly 5m (16ft) of fill had been added over previous decades. This sat on top of a layer of hard clay of variable depth, below which was soft clay prone to oozing. The limestone bedrock was as far as 30m (98ft) down.

Buildings constructed in the immediate aftermath of the Great Fire were mostly four- to five-storey load-bearing masonry structures. These were founded on the hard clay 5m (16ft) down, usually supported by a rubble stone mat. Unfortunately, the soft clay below often failed to offer long-term support, and many buildings began to sink. Luckily, this settlement was usually accommodated, but in extreme cases it could be as much as 600mm (24in), which caused significant structural damage.

Chicago's new post-fire City Hall took ten years to build and featured an experimental

BELOW The first office block with a riveted all-steel frame was the ten-storey Rand McNally Building, opened in 1889. *(Public domain)*

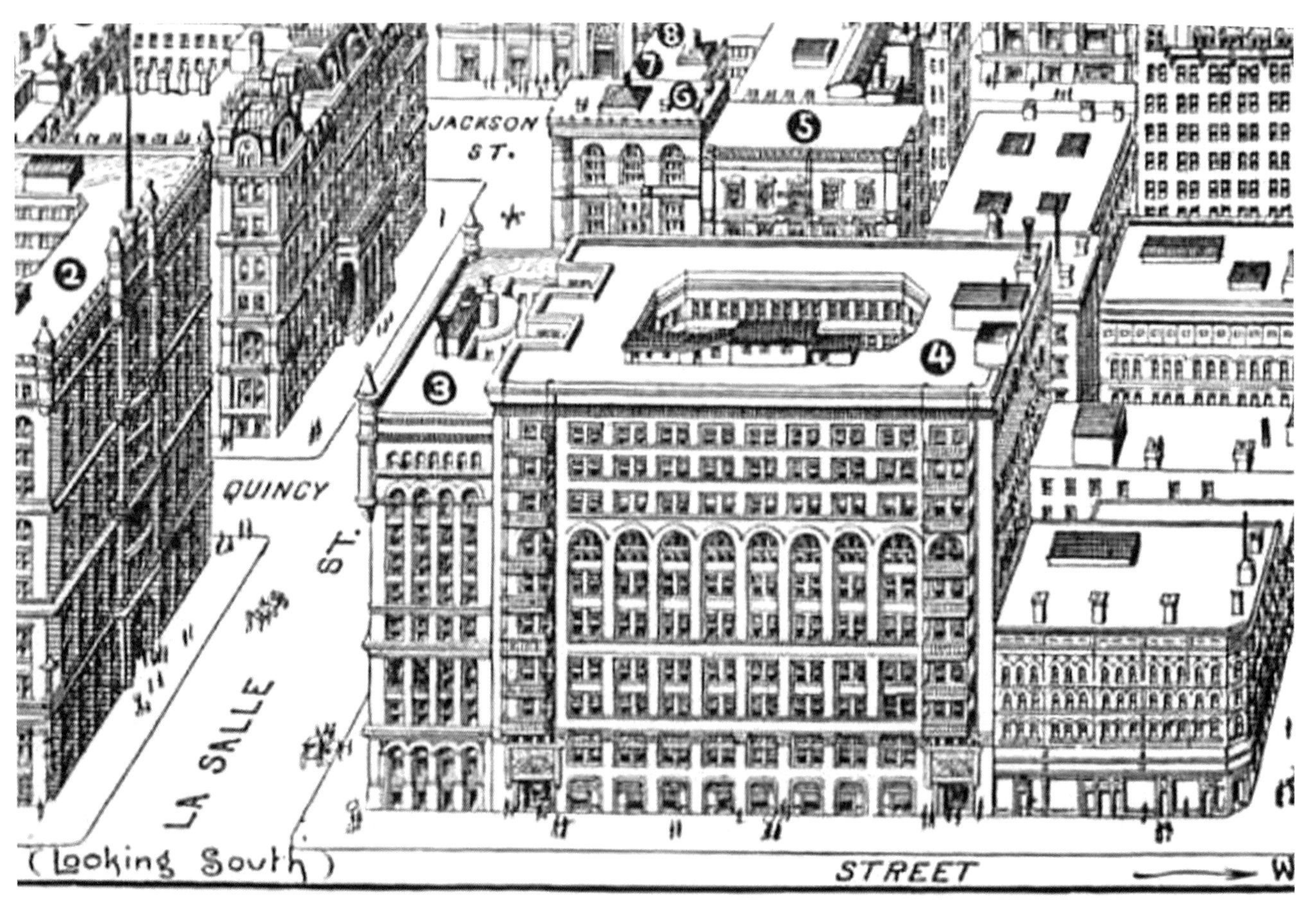

mat and pile foundation. By 1905 the building had sunk 150mm (6in), a gas pipe fractured and the resulting gas explosion blew the roof off. It was demolished and replaced by a steel-framed building that is still in use today.

By the 1880s many of the low-rise buildings erected in the immediate aftermath of the fire were already being demolished to make way for significantly taller structures. Two key innovations, both pioneered in the UK, came together to make it practicable for Chicago designers to look upwards.

Pneumatic caissons enabled builders to excavate 30m (98ft) down to the solid bedrock – well below the water table – and install settlement-free foundations. And the increasing availability of low-cost Bessemer steel sections meant that metal framing was both cheaper and more reliable.

Often dubbed 'the first true skyscraper', the 42m (138ft), ten-storey Home Insurance building built in 1884 was in reality more of a transitional design incorporating elements of both old and new construction methods. Its structural frame used wrought iron alongside steel, and load-bearing masonry also featured. A more modern building opened in 1889. The 45m (148ft) ten-storey Rand McNally Building is notable for its all-steel frame, riveted together rather than bolted.

These early taller buildings had to rely on town gas for internal lighting. Bright electric carbon arc lamps became available at the start of the 1880s, but they flickered and hissed and required frequent manual adjustment. Incandescent bulbs were still in their infancy. They generally used carbon filaments, which led to the gradual build-up of carbon deposits on the inside of the glass bulb. It was not until well into the next century that the crucial development of tungsten filaments created a long-lasting light bulb.

However, as buildings up to 58m (190ft) tall sprang up in the city centre, opposition to the whole concept was hardening. These new buildings usually had coal-fired boilers in their basements supplying steam for heating, ventilation and electricity generation – and probably elevators as well. The coal smoke they belched out seriously polluted the air in the streets they overshadowed. In 1892, Chicago banned buildings over 46m (150ft) tall.

Reaching for the sky

There were no height restrictions in New York in the 19th century. The city had grown up on the island of Manhattan, and in 1835 the last of a series of devastating fires destroyed hundreds of wooden buildings but only claimed two lives. As in Chicago decades later, low-rise masonry buildings were the obvious replacements. Manhattan is approximately 22km (13.4 miles) long and less than 4km (2.3 miles) at its widest point, and the city had initially grown up at its southern tip. As the city boomed in the 19th century, immigrants flooded in from Europe and beyond, and it began to spread rapidly northwards.

New York soon became a major financial centre. The demand for ever more prestigious office space grew exponentially. Nevertheless, the city was slower to adopt the skyscraper concept pioneered in Chicago, not least because the city authorities were reluctant to approve metal-framed buildings of any height. This began to change in 1888, when plans for the 11-storey Tower Building were approved.

Again this was a transitional design that used cast-iron columns. A fine example of a skyscraper in the modern idiom was the 21-storey, 103m (338ft) tall American Surety Building. Opened in 1896 it topped anything in Chicago and triggered a long-term obsession with height record-breaking in New York.

The skyscraper boom that followed was made possible by a less well-known development in Chicago.

As buildings there grew taller, it became increasingly obvious that architects' training did not equip them to design ever more highly stressed structures. A new professional discipline emerged – that of the specialist structural engineer. Added to the earlier developments of Bessemer steel, pneumatic caisson foundations and the safety elevator, this new expertise meant that in practice the architect could propose ever-taller skyscrapers with the confidence that they would stay up.

Manhattan's skyscrapers soon began to break height records. In the first decades of the 20th century several buildings in close succession laid claim to be the world's tallest. In 1909 the Metropolitan Life Insurance Building passed the 200m (656ft) mark, only

OVERLEAF

Manhattan island in 1873. *(Public domain)*

Entered according to act of Congress in the year 18
PUBLISHED BY GEO. DEGEN, 22 BEEKMAN ST. N.Y.
NEW

n the Office of the Librarian of Congress at Washington, D.C.
PRINT. BY G. SCHLEGEL, 97 WILLIAM ST. N.Y.
ORK,

to be topped four years later by the 241m (792ft), 60-storey Woolworth Building. This held the record until 1930.

Not every citizen of New York shared the property developers' enthusiasm for tall towers. Typical skyscrapers of the early 20th century rose straight up from the sidewalks, turning streets into dark canyons where the sun never penetrated. There were concerns that New York's firefighters would be unable to cope with a skyscraper fire. It was even alleged that the Great Baltimore Fire of 1904, in which more than 1,500 buildings were destroyed but no lives were lost, was made worse by the (few) high-rise buildings in the city.

Baltimore quickly banned buildings more than 21m (70ft) tall, but campaigns to impose specific controls on skyscrapers in Manhattan dragged on for decades. Eventually, complex new zoning laws were introduced in 1916. Buildings could only rise 30m (98ft) from the sidewalk, before setting back at a prescribed angle over 75% of the building plot to allow sunlight to reach the streets below. These stepped exteriors gave New York skyscrapers their distinctive style.

However, developers had a free hand over 25% of the plot and could build as high as they liked. From the end of the First World War there was a skyscraper boom in New York that led eventually to the 1931 construction of the long-time height record holder, the iconic 381m (1,250ft) tall Empire State Building.

Chicago was also booming, and eventually the civic authorities had no choice but to lift most building height restrictions. The New York setback style never caught on in Chicago, which developed its own style in response to the complex local regulations. Skyscrapers began to appear in other American cities – and in other countries – but never in the numbers that sprang up in Chicago and Manhattan.

These two cities built tall for good reasons – a shortage of desirable building land, unprecedented economic growth and a rocketing demand for modern office space. Elsewhere at that time it was much harder to justify a skyscraper, and those that did appear were status symbols more than economically viable projects. It was 50 years or more later that skyscrapers began to appear in significant numbers in major cities across the globe – and many of them were vanity projects.

OPPOSITE Built in the classic New York setback style, the 241m (792ft)-tall Woolworth Building was the world's tallest from 1913 to 1930. *(Aude)*

THE MAN WHO PATENTED THE SKYSCRAPER

In May 1888 a certain LeRoy Sunderland Buffington was granted a patent covering the construction of tall buildings with load-bearing iron frames – the key element in the early development of skyscrapers. Buffington, an architect from Minnesota, USA, claimed to have invented the system in 1881 – four years before the first such building rose in Chicago.

Nearly all subsequent 19th- and early 20th-century skyscrapers used the same basic system – but Buffington received little in the way of acknowledgement or financial benefit from their developers.

His requests for patent royalties were either ignored or ridiculed, with only one significant exception, the 95m (311ft) 26-storey Rand Tower in Minneapolis, completed in 1929. This was just two years before he died at the age of 84, still in private practice as an architect.

LEFT LeRoy Sunderland Buffington may have patented the skyscraper, but he was almost totally ignored by skyscraper builders. *(Public domain)*

Chapter Two

Living with skyscrapers

Love them or loathe them, skyscrapers are now a normal feature of the urban scene. This chapter examines the impact of skyscrapers on the city and in popular media, as well as considering what it is really like to live so high in the sky. Some skyscrapers are seriously quirky – we look here at some of the more amazing.

OPPOSITE One of the most iconic and reproduced images, 'Lunch Atop a Skyscraper' immediately conjures up a sense of height without actually featuring the Art Deco skyscraper the ironworkers were building. *(Public domain)*

Skyscrapers in popular culture

Some of the world's most famous skyscrapers have earned notoriety through appearances in photography, adverts, films and other works of fiction.

From the early days of prolific high-rise building in the USA, many a photographer and filmmaker felt the appeal of using skyscrapers as the perfect victim in any disaster film, and for providing the striking backdrop for a death-defying stunt.

The iconic 1932 black and white photograph entitled 'Lunch Atop a Skyscraper' is one of the most reproduced, and even parodied, images known worldwide. Arguably it is the most evocative skyscraper image out there, yet the building is in fact an unseen part of the image. Depicted are 11 ironworkers sitting side by side on a steel beam eating lunch with their feet dangling freely from the 69th floor of the Art Deco building that formed part of the Rockefeller Center in Manhattan.

As natural as they look, the men were in fact part of a publicity shoot to create a hubbub around the soon-to-open building. The image remains one of the most evocative in capturing the feel of being atop a skyscraper during its construction.

On film, silent movie star Harold Lloyd created one of the most enduring skyscraper stunts that pre-dated similarly famous high-rise exploits by decades. Try as you might, it is near enough impossible to unsee the frightening yet funny scenes featuring the hijinks of Lloyd's alter-ego as he scales a tall building in the 1923 film *Safety Last!* Even if you have never watched the full film, there is every chance you will have seen the iconic still of the character's well-dressed form dangling precariously from a clock face: just one of the many amusing but excruciating hazards he meets during his nail-biting ascent.

The film is a perfect early example of that unofficial genre of filmmaking that compelled many an audience member to watch agonising scenes through splayed fingers.

BELOW **The building that the 11 ironworkers, depicted in 'Lunch Atop a Skyscraper', were working on during the photoshoot.** *(Shutterstock)*

LEFT Harold Lloyd dangling precariously from the clock face of a skyscraper in the 1923 silent film *Safety Last!* is one of the earliest and most enduring appearances of a skyscraper in popular culture. *(Public domain)*

Richard Pryor gives a comedic nod to Lloyd during his skyscraper scene in the 1983 superhero film, *Superman III*. In this case the unnamed building serves as the backdrop for a vertiginous ski slope. Pryor's character attempts to describe the Man of Steel's escapades with a pink tablecloth draped around his shoulders and skis strapped on his feet. Unfortunately, he loses his balance and accidentally skis off the top of the skyscraper's snowy slope, drops down on to an inclined glass roof and somewhat improbably lands perfectly safely on the road below, before nonchalantly shuffling off through the traffic.

A much better-known skyscraper icon featured as the backdrop for a more elaborate and contemporary stunt in the film *Mission: Impossible – Ghost Protocol* (2011). Tom Cruise, known for performing many of his own stunts, plays the role of spy, and apparent adrenaline junkie, Ethan Hunt. In one startling scene, Cruise/Hunt, with the aid of some sturdy safety harnessing, propels himself out of one of the windows high up the world's tallest building, the Burj Khalifa. Not content with a simple scene-stealer clinging to the side of the 828m (2,717ft)-tall monolith, Cruise's character produces some extraordinary acrobatics, including running along and around the building's façade with apparent ease.

Almost as free and easy, but more death-defying moves are the focus of the carefully and intricately planned performance that is detailed in the 2008 film *Man on Wire*. This is not a fictional construct but rather is the true story of the great tightrope-walking prowess of Philippe Petit performing a sensational feat of walking on a high wire strung between the tops of the Twin Towers at New York's World Trade Center in 1974. The film brings together stunning and real footage of the high-wire artist and the years of planning that saw him perform illegal – yet eye-popping and death-defyingly artistic – moves along the 400m (1,312ft) high wire for almost an hour.

LEFT The infamous former World Trade Center Twin Towers provided support for the sensational high-wire performance artist Philippe Petit in 1974, as featured in the 2008 documentary *Man on Wire*. *(Shutterstock)*

ABOVE The helipad on the Burj Al Arab has acted as a makeshift tennis court for Andre Agassi and Roger Federer, and a track for a Formula 1 wheelspin trick – both headline-grabbing promotional stunts. *(Shutterstock)*

The imagination of stunt designers has continued to be sparked by these tall architectural monuments. In recent years, the Burj Al Arab's 24m (79ft) wide helipad, which projects out of the top of the Dubai building's sail-like form, has provided a platform for some of sport's biggest names to take part in similar promotional tricks. Not least of these are the doughnut wheel spins performed by Formula 1 driver David Coulthard and a casual knockabout on a makeshift tennis court between Andre Agassi and Roger Federer – all high up on the 321m (1,053ft)-high skyscraper.

Back in the world of make-believe, Hollywood's fascination with tall buildings has created more sinister affiliations. One of its most iconic stars, Steve McQueen, helped draw the audiences to see his character attempt to rescue partygoers from a terrifying fire at a recently opened, but doomed, fictional tall office building in the 1974 epic *The Towering Inferno*. The aptly titled *Skyscraper* (2018) action flick pays homage to the McQueen film featuring a security expert protagonist seeking to save his family and other innocents from a fire in the world's fictional tallest skyscraper in Hong Kong.

With terrorists playing a pivotal role in the plot, the film also clearly references mystery thriller *Die Hard* (1988), in which Bruce Willis's character battles with terrorists to free hostages in a Los Angeles skyscraper

This Los Angeles skyscraper served as the fictional Nakatomi Plaza featured in the film *Die Hard*. *(Shutterstock)*

and performs stunts such as the memorable moment he leaps, with a makeshift harness, from the building as it explodes all around him.

Destruction of these supertall icons is a motif that has been repeated often over the years. From the most famous monster movie, *King Kong* (1933) featuring the legendary giant ape swiping at attacking planes while clinging to the top of the Empire State Building, to the 1998 remake of that other iconic monster movie, *Godzilla*, the filmmakers have turned to skyscrapers again and again. *Godzilla* takes inspiration from Manhattan's skyline and features key appearances by the Chrysler and MetLife buildings amid the chaos. The New York skyline is so iconic it has provided the inspiration for many science fiction cinematographers, not least for the dystopian backdrop to *Blade Runner* (1982) and the 1927 pioneering German silent film *Metropolis*.

Other sci-fi movies have embraced the idea of skyscraper icons as collateral damage – the former Twin Towers often the victims here, being subjected to devastating meteor showers in *Armageddon* (1998), and a comet-induced giant wave in *Deep Impact* (1998). In a variation on the theme, Chicago's 343.7m (1,128ft) tall John Hancock Center tower features as the spooky site for the supernatural horror film *Poltergeist III* (1988).

And, of course, elsewhere the skyscraper features in a more sympathetic light – whether as the glamorous Tokyo Park Hyatt hotel bar backdrop in *Lost in Translation* (2003), or the Empire State Building providing the romantic meeting point for star-crossed lovers in *Sleepless in Seattle* (1993).

The opening scenes of *Manhattan* (1979) are perhaps some of the most romantic of all. With the images of the bustling city and iconic black and white skyline set to the sounds of George Gershwin's 'Rhapsody in Blue', this famous scene is a love letter to the most famous skyscraper city in the world.

BELOW LEFT One of the earliest monster films sees King Kong scale the full height of the then recently opened Empire State Building, the world's tallest at the time, in the 1933 film. *(Shutterstock)*

BELOW The John Hancock Center tower in Chicago provided the spooky backdrop for supernatural horror *Poltergeist III*. *(Shutterstock)*

When skyscrapers go wrong

Advances in engineering over the decades of skyscraper design have made once-impossible heights now unremarkable. Sometimes, however, pushing the engineering and architectural boundaries means learning on the job what is and what is not possible.

Engineering out risk is key, and sometimes new risks present themselves at different stages of project. In 2014, a striking new tall building at 20 Fenchurch Street in London, which would come to be known as the 'Walkie Talkie' thanks to its unusual profile, was almost complete when it emerged the building was having a strange and surprising effect at ground level. Designers had to incorporate a new so-called sunshade feature high up the structure because the curved glazed façade was causing a solar glare so extreme the heat had damaged a car parked on the street below.

Just a few years earlier, and a little south of the Thames, the green ambitions of the Strata SE1 tower at Elephant and Castle in South London were somewhat stymied as it emerged that the three turbines installed quirkily at its apex were not in fact generating the hoped-for wind power benefits to residents and occupants. Some reports have suggested the turbines have not turned effectively at all since the building's opening in 2010.

A few decades earlier, there was the staggering story of Manhattan's Citicorp (now Citigroup) Center, completed in 1977. This stately icon stands out on the infamous skyline thanks to its dazzling white gleam and 45° pitched roof. Even at ground level the 59-storey building is a marvel with its four 9-storey-tall column supports located at the centre of each side, rather than the more regular corner positioning.

This design feature was a response to planning requirements and the constraints they

LEFT Designers had to add a sunshade feature to the still-under-construction 'Walkie Talkie' in London after solar glare from its curved façade caused damage to a car parked below. *(Shutterstock)*

LEFT Despite the three wind turbines being a key feature of the Strata SE1 tower, there is little evidence they have worked effectively since the building opened. *(Shutterstock)*

imposed because there were plans for a new church to be built on the adjacent site.

As a result, above these mid-side columns, the Citicorp building cantilevers out 22m (72ft) to avoid encroaching on the new church next door and creates a plaza with impact at ground level.

Its looks were not the only thing that set this 279m (915ft) tall building at 601 Lexington Avenue apart. Among its clever design elements, overseen by structural engineer William LeMessurier, were the addition of early energy efficiency measures not widely used at

BELOW The stately white gleam and pitched roof of Manhattan's Citicorp Center is an icon on the Manhattan skyline. *(Shutterstock)*

the time and the development of the first tuned mass damper in a tall building in the USA. This almost 372t (410 US tons) block of concrete is positioned on an oiled plate at the building's peak and behaves like a pendulum, reducing sway by moving out of phase with the building as it responds to the wind (*see* Standing tall – the structural frame, p.82).

The building hid an extraordinary secret for many years, one that seems almost fictional, but that ultimately tells the tale of the ethics of engineering when an engineering success threatens to turn into an engineering nightmare.

In the summer of 1978, following an academic query relating to the building's performance under wind loading, LeMessurier returned to investigate the tower's unusual structure. Above the four columns at the base, the building is supported with stacked load-bearing steel cross-braces arranged in a chevron pattern. These were designed to brace the building against winds hitting the façade in a perpendicular motion. But quartering winds, which hit at the corner and affect two sides at once, began to cause concern following a recent revelation.

LeMessurier had discovered that the steel connections had not in fact been built exactly as specified, and instead of being welded at key locations in the bracing, had been substituted with more cost-effective, and less robust, bolted connections.

Substitutions such as these are often made as a project is refined and are perfectly safe, and accurately assessed to be so, in most instances. But in this case, this knowledge was coupled with LeMessurier's growing concern that quartering winds had not been accurately accounted for during the design. LeMessurier faced a huge dilemma on whether to raise the alarm and risk retribution. He commissioned experts and wind tunnel testing that confirmed, and exacerbated, his fears.

The most vulnerable connections, according to his revised calculations, meant the building was at risk from the winds of a storm that might happen once every 16 years – potentially far too frequent an event to give reassurance the building would not be in danger of a catastrophic failure.

An extraordinary plan was hatched to

access the original bolted joints and reinforce them with new steel plate patches welded over the weaknesses. Time was against the remediation work as the start of the hurricane season loomed. In September 1978, as work continued apace, Hurricane Ella gave the team cause for great concern as it made its way towards the city as the most vulnerable connections were being rapidly repaired.

With just hours to spare until it made landfall, Ella shifted its path and began moving away from Manhattan, much to the relief of LeMessurier who had no desire to see the building's almost-complete repairs tested so dramatically.

Welding work was completed by October, making the building one of the most structurally robust towers in New York.

Legally and financially there were consequences for LeMessurier and the project team involved. The story remained largely untold until a revelatory article was published by the *New Yorker* magazine in 1995, but in engineering circles, the actions of LeMessurier are a reminder of how important the ethics of engineering and architecture are to the safety of a building's inhabitants.

Skyscrapers and the city

Chicago is usually regarded as the birthplace of the skyscraper – yet in 1892 the city banned the construction of any more skyscrapers, then defined as any building more than 45.7m (150ft) tall. New York imposed very tight new regulations on tall buildings in 1916 (*see* The origins of the modern skyscraper, p.26). The new generation of buildings more than ten storeys tall aroused very negative feelings among many New York and Chicago residents. There was organised resistance against almost every tall building proposal. But why?

More than a century later that negativity and resistance still exists, particularly in Western cities that developed before the arrival of large-scale car traffic. In Chicago and New York, the original objections were founded on the impact of the taller buildings at street level. The choking fumes from the coal-fired boilers in every 19th-century tall building's basement may no longer threaten pedestrians' lungs, but it is still the impact of a skyscraper on the surrounding streets that is one of the primary reasons for resistance, even in the 21st century.

Some cities simply ban skyscrapers from their historic centres, banishing them to out-of-town locations. The French capital, Paris, is a prime example. The historic centre, a major tourist destination, remains virtually unchanged from its 19th-century heyday. Skyscrapers can only be found 3km (1.9 miles) to the west of the city limits, on the 560ha La Défense business district, which dates from the 1960s. Initially tall buildings here were restricted to just 100m (328ft), although this limit was lifted in the 1970s.

London had a similar policy for a while. The city had an ambivalent relationship with skyscrapers in the 1980s. Until 1991 there were only two true skyscrapers in London, Centre Point (*see* Centre Point Tower, 1966, p.135) and what is now known as Tower 42. Both highly controversial, they did little to advance the case for skyscrapers in the city.

The derelict London Docklands to the east of the city were seen simultaneously as an eyesore and a great development opportunity.

OPPOSITE The Citicorp Center's distinctive design was found to be fatally flawed after its opening and the building underwent secret structural repairs to remedy the problems. *(Shutterstock)*

BELOW Paris keeps skyscrapers well away from its historic city centre. *(David McSpadden)*

Tower 42 was one of only two skyscrapers in London until 1991. *(Diego Delso)*

So it seemed logical to follow Paris's example and create a business district where skyscrapers could flourish.

Communications were the key. The Isle of Dogs, the River Thames peninsula where the first docks were built, had always had poor transport connections. To spur commercial development the new London Docklands Development Corporation sanctioned the construction of the driverless Docklands Light Railway. This was a modest investment, as the original phase of the line ran along disused railway infrastructure and across derelict land.

An improved and very expensive road link to the main east London road network was also promised. This was enough to get commercial developers on board, despite the government originally refusing to fund a link to the London Underground network. A second major financial centre began to take shape, epitomised by the 236m (774ft) One Canada Square tower on Canary Wharf. Opened in 1991, it surpassed Tower 42's 183m (600ft) as the UK's tallest building, a record it held until The Shard, at 310m (1,076ft), topped it 20 years later.

London had in fact been the second major city to introduce height limits for city-centre buildings. In 1894 the London Building Act banned buildings more than 24.4m (80ft) tall. This was said to have been in response to the construction of the 14-storey 49m (160ft) residential Queen Anne's Mansions in Westminster, which attracted many complaints at the time, not least from Queen Victoria, who could no longer see the Palace of Westminster buildings from her window in Buckingham Palace.

Taller buildings arrived in the 1960s with the lifting of the height restrictions. There was no immediate boom in skyscraper construction, partly because of strict regulatory limits to protect sight lines to St Paul's Cathedral, the Palace of Westminster and the Tower of London. Other historic cities have similar safeguards. It was only in the first decades of the 21st century that skyscrapers began to sprout in central London – and Londoners began to experience the pluses and minuses of these shiny new towers, something that the inhabitants of Manhattan in particular had known for a considerable time.

During working hours, office skyscrapers can house thousands of staff. All of them have to get into work every morning and get home every evening. Many such skyscrapers have multi-level basement parking, encouraging staff to come to work by car. In the UK it used to be common practice for service companies to give many of their staff company cars as well as guaranteed parking spaces in the basement, rather than subsidising rail season tickets or similar.

So simply planting a skyscraper into a city-centre site usually had the predictable result of significantly increasing road traffic congestion in the area, not to mention the associated air pollution. There are many cities around the world now where the morning and evening commutes by car can take several hours just to travel a relatively short distance between the residential areas on the fringes of the city and the serried ranks of office towers in the centre. Most of these cities inevitably suffer from health-threatening emissions as well.

Actively discouraging staff from commuting in cars is not a comprehensive answer. Major cities tend to have public transport infrastructure developed over decades. Significantly increasing the capacity of underground rail systems or introducing innovative light rail networks can take decades more. A rapid boom in office skyscraper construction can rapidly overwhelm existing capacity.

Back in the 1950s railroad companies in the USA realised that their tracks into city-centre stations passed through land that had become extremely valuable as the cities boomed. To take advantage, platforms were constructed above the tracks and the so-called air rights sold to commercial developers. There were even plans to replace New York's iconic Grand Central Terminal with a 50-storey tower. Thankfully, public protests saved the station itself, but the massive 246.3m (808ft) 59-storey Pan Am Building (now the MetLife Building) rose next door, to be itself the focus of much criticism for its size and its blocking of cherished city views.

ABOVE LEFT London's Docklands were transformed in the 1990s, with the One Canada Square Tower becoming its 236m (774ft) centrepiece. *(Mattbuck)*

ABOVE First known as the Pan Am Building, the MetLife Building was slammed for blocking cherished Manhattan views. *(Postdlf)*

Locations close to major transport hubs have always been the skyscraper developers' favourites. Air rights buildings rose wherever a suitable opportunity presented itself. In London, the Broadgate development in the 1990s sat above the tracks approaching Liverpool Street Station, and over part of the station platforms as well. So far nothing over 35 storeys has been proposed for an air rights tower, mostly because of the inevitable complexity of the foundations.

At street level, pedestrians can be subjected to some uncomfortable and even dangerous side effects of skyscraper construction. Every high-density city generates its own comfort blanket of warm air held in place by the denser cold air above. This warmth comes mainly from buildings, roads and paved areas radiating solar energy back into the atmosphere, although heat leaking from poorly insulated buildings and vehicle emissions also play a part. Skyscrapers, however, rise through the blanket and into the cold air above.

When the winds are right, this cold air hits the upper-level windward face of the skyscraper – and some flows down rather than sideways, all the way to the base. Down on the street, pedestrians will feel an unpleasant chilling, something of a shock if they have started their journeys in low-rise areas of the city where the comfort blanket is undisturbed.

RIGHT Bridgewater Place tower in Leeds is notorious for the fierce gusts of winds it generates at street level. *(Lad 2011)*

Models of skyscrapers have been tested in wind tunnels for many years now – but almost always in isolation and purely with the objective of minimising the wind loading and sway. Little attention was paid to the possible impact of winds at the base, especially where skyscrapers are clustered together. But even a single skyscraper can trigger dangerous gusts at street level.

Perhaps the most notorious case is that of the 32-storey, 112m (367ft) Bridgewater Place tower in Leeds, England. Since it opened in 2007 pedestrians have been blown over and injured, baby buggies have been swept into the street and in 2011 a man was killed when a lorry overturned on top of him. Authorities reacted by closing the streets around the tower whenever wind speed exceeded 72kph (45mph). Remedial measures proposed included retrofitting some form of external aerodynamic deflectors.

It was only when reports like these became impossible to ignore that the regulatory authorities began to clamp down. This was at a time when the computing capacity available to building designers had massively increased. A new technique, computational fluid dynamics (CFD), made it possible for the complex airflow patterns around tall buildings to be analysed in astonishing detail. CFD could also accurately model airflow around a cluster of buildings, both tall and medium-rise, and identify regions of the streets where dangerous gusts might strike.

In 2019 the City of London Corporation brought in new guidelines that require skyscraper developers to undertake detailed CFD analyses and wind tunnel testing on all proposed towers early in the design process. The object is to ensure as much as possible that the skyscraper, as built, will minimise the risk of dangerous gusts occurring at ground level, rather than discovering that what is erected needs urgent modification with external aerodynamic devices, such as in the case of Bridgewater Place.

Ideally, skyscraper designers will also take into account the other impacts a new development might have on the urban environment. Skyscrapers can seem like alien invaders squatting in the city, with little or no reference to the historical development nearby and minimal consideration for the pedestrians that have to coexist with them. London's notorious 117.3m (385ft)-tall Centre Point tower, for example, encroached onto the original pavements around it. When these narrower pavements became crowded, pedestrians were frequently forced out into the busy road alongside, at risk of injury or even death.

It was the 157m (515ft) Seagram Building in Manhattan that first showed there was a better, more sympathetic approach (*see* Seagram Building, 1958, p.131). It occupies only part of its (very expensive) plot on New York's prestigious Park Avenue with a granite public plaza in front of it. This is one tall building that is universally popular with New Yorkers and a lesson for all skyscraper developers all over the world.

Life in the sky

For many decades, skyscrapers were simply giant office blocks, perhaps with ground-floor retail space, but little else in the way of amenities. Commercial pressures dominated design. Maximising the floor area available to let was the overriding priority for architects and structural engineers alike. Along the way, the impact of such profit-driven accommodation on the hordes of workers crammed into identical office floors was almost totally ignored.

Towards the end of the 20th century things began to change. Greenery began to sprout in unlikely locations high up above ground level. Atriums, basically internal spaces reaching up many floors, became popular features. Companies seeking new office space began to realise that the better the design of the skyscraper, the more productive would the staff be and the quality of applicants for any position would be higher. Mixed-use buildings became much more common, presenting novel challenges to the designers. The 21st century would see a new breed of tall buildings – easier to live in, easier to live with and significantly more effective as a result.

One of the first skyscrapers to challenge the status quo opened in 1997. Architect Norman Foster and structural engineering firm Arup were given the same free hand granted to the designers of New York's Seagram Building

Frankfurt's Commerzbank tower broke new ground in skyscraper design. *(Mylius)*

ABOVE A sky court on the 19th floor of the Commerzbank. *(Wolfgang Meinhart)*

and London's the 'Gherkin'. Their brief was to design a new landmark headquarters building in Frankfurt for Germany's Commerzbank, and the result was dubbed the world's first ecological skyscraper.

An antenna spire takes the overall height of the 53-storey building to about 300m (984ft), although roof level is 258.7m (849ft). The floor plan is unique. A full-height central triangular atrium forms the 'stem', around which are grouped three linked stacks of floor plates forming the 'petals'. Sealed four-storey 'sky courts' – essentially internal atriums – spiral around the central atrium and are accessible from distinct four-storey groups of offices.

Each sky court is heavily planted with shrubs and trees. Foster envisaged the tower as a community of four-storey 'villages', each with around 240 inhabitants, and with the sky courts acting as village greens. There are coffee bars and seating tucked away among the planting, encouraging the inhabitants to meet, socialise and exchange ideas.

Sky courts are becoming increasingly popular, but it has to be said that some versions have been less than successful. Large external balconies, regardless of how thoughtfully planted they are, have often suffered badly from the high winds frequently encountered at height. The same applies to rooftop gardens – or sky gardens – although designers have gradually learned to create effective wind deflectors and shelters.

OPPOSITE The Shard put on a spectacular light show at its inauguration. *(Richard Goldschmidt)*

Many otherwise private tall buildings now allow public access to their sky gardens – for a fee, of course, in most cases. There can be found bars, restaurants, even swimming pools, all with fabulous views. It is worth noting that in the 1930s the Empire State Building in New York was only saved from bankruptcy by the admission fees paid by hordes of tourists flocking to its 102nd-floor viewing platform, and many supertall buildings since have followed its example.

One of these is London's 310m (1,017ft) The Shard. This has a viewing platform with three floor levels at its peak, highly popular with tourists prepared to shell out a substantial entry fee. In fact the 95-storey tapering tower, opened in 2013, is a prime example of the mixed-use supertall concept that has become increasingly popular this century.

Of the 72 habitable floors, the first 28 are devoted to high-end office space. A 200-room, five-star hotel occupies the 34th to 52nd floors, with luxury residential accommodation from the 53rd to the 65th. A mechanical services floor separates each zone. Between the offices and the hotel, however, is a three-storey sky court. This is intended as a social space, with shops, restaurants, bars and exhibition areas. Architect Renzo Piano envisaged The Shard as a 'small, vertical town for about 7,000 people to work in and enjoy and for hundreds of thousands more to visit'.

This concept of the 'vertical village', or even the 'city in the sky', has intrigued designers for many years. A mixed-use supertall skyscraper sitting astride a major transport hub would be the closest to a vertical city that is likely to be achieved in the near future. Residents and workers would have little need to descend to street level, except to make out-of-town journeys. In the sky courts and roof gardens they would find all the facilities and services they would need on a daily basis.

There would be supermarkets, artisan bakers, hairdressers, even cinemas. If the building is large enough, there could be doctors' surgeries, dentists and opticians. Kindergartens might be popular, along with libraries and performance spaces. Almost certainly some sky courts would be given over to gyms, swimming pools and saunas. Luxuriant planting would be universal. And there would inevitably be a wide choice of restaurant cuisines, just as there is in The Shard.

There was a time when every building relied on its external walls to carry floor and roof loads as well as protecting its occupants against the elements. These load-bearing walls had one major limitation – the higher they were built, the thicker the walls had to be at ground level (*see* The origins of the modern skyscraper, p.26). Ten storeys was the practical limit, much lower if the ground conditions were less than favourable.

A different approach was needed, and in Chicago in the late 19th century the solution

RIGHT Lever House is where architects fell in love with the all-glass curtain wall. *(Beyond My Ken)*

emerged from earlier technical developments such as cheap steel and the safety elevator (*see* The three key developments, p.19). The structural and weatherproofing functions were separated. An internal steel skeleton supported all the floor and roof loads, while a non-structural curtain wall hung off the framework kept the weather out.

In the early days, architects sought to reassure those wary of much taller buildings by making up the curtain wall from steel-framed cladding panels that featured traditional walling materials such as stone or terracotta. Windows were little larger than in traditional construction.

This approach continued for decades. Then, in the 1950s, American architects fell in love with glass. Developments in glass technology meant that designers could dream of a totally transparent building, clad in nothing but high-performance glass panels. The first such arrived in 1952 – the 24-storey Lever House in Manhattan. Its 93.6m (307ft) height may have been modest, even by the standards of the 1950s, but its influence on skyscraper design was profound.

In practice glass façades turned out to be something of a mixed blessing. Yes, natural daylight could penetrate deep into the floor plan – but in summer the solar heating of what was essentially a giant multi-storey greenhouse had to be combated by a massive energy-hungry air-conditioning system. In winter the heating bills were colossal, the glass having little insulating value.

Architects may have lauded it, visitors were impressed. Occupants were less enthusiastic. Those with even the slightest touch of acrophobia hated working close to the external glasswork, and almost all complained about the lack of individual control of heating and ventilation.

Glass technology continued to improve, however. Glass became stronger and safer. Double-glazed panels helped to cut heating bills. Solar gain could be reduced by a wide variety of tints and surface treatments, including coatings of real gold. Occupants could be given more control of their immediate environment, although opening windows was not a realistic option on upper floors due to the high winds common at those levels.

Curtain walls became more complex. By the 21st century as much engineering innovation was going into the cladding as into the rest of the building. So-called active cladding emerged, usually featuring arrays of sunshades that react automatically to variations in the intensity of sunlight, making all-glass curtain walls much more acceptable to occupants. Translucent photovoltaic (PV) cells featured on many façades. And the introduction of 'double-skin' cladding made occupant-controlled natural ventilation possible once more.

Not all skyscrapers were glass clad. Designers could opt for a mixture of glass windows and decorative infill panels. These could be 75–100mm (3–3.9in)-thick slices of marble, limestone or granite, metals such as zinc, copper, stainless steel and even bronze, or some form of insulated composite panels.

Modern curtain walling systems are universally framed in extruded stainless steel or aluminium sections, some of them large and complex. Aluminium's performance in fire leaves much to be desired, however. Elaborate measures must be taken to minimise the spread of fire upwards in the gaps between the floor slabs and the curtain wall. In practice, the only effective method of ensuring that a fire never leads to the catastrophic failure of aluminium-framed curtain walling is the fitting of automatic sprinklers.

Cladding panels usually span more than two floors, although their maximum size is basically limited by the logistics of hoisting them into place. These days, panels will be manufactured offsite in factory conditions, so the design has to be able to be delivered to the site by road transport. Wind is another constraint. Large, lightweight panels can only be lifted externally to the building in light winds.

Today's cladding designers have another priority. The curtain wall has a major effect on the building's carbon footprint. It has to be very well insulated and able to minimise the effects of solar gain in the summer, while being able to take maximum advantage of it in winter. This is easier than it once might have been following the introduction of even higher-performance glass and even triple-glazing.

Another new concept is that of the skyscraper cluster. Two or more identical towers are built in close proximity and linked together by sky bridges at one or more levels. The 452m (1,483ft) Petronas Towers in Malaysia, once the tallest buildings in the world, were the pioneers of this concept (*see* Petronas Towers, 1998, p.139). Sky bridges are usually fully enclosed, providing not just popular social spaces but alternative escape routes in emergencies.

Modern designers may not realise it, but the concept of sky bridges dates back to at least the 17th century, as they were a common feature in the Yemeni walled city of Shibam (*see* Ancient 'skyscrapers', p.12). The Marina Bay Sands complex in Singapore takes the concept one stage further – a single 340m (1,120m)-long sky garden tops three identical 190m (623ft) high towers. There are the usual bars and seven 'celebrity chef' restaurants, an actual garden and a 150m (490ft)-long infinity swimming pool.

Naturally there is a hefty entrance fee, hardly surprising as the complex is in reality a resort hotel with a giant casino at its heart, much like the glitzy casinos in Las Vegas. The Shard is a much more relevant example for future skyscraper developments.

Apart from social spaces, lush greenery and far-reaching views, what should residents and office workers expect from the mixed-use skyscraper of the future? The images of 9/11 are seared into the public memory; so top of the list will be safety and security.

Lessons have been learned from the 9/11 disaster (*see* Lessons from 9/11, p.177). Advanced automatic sprinkler systems will be central to the fire-control strategies of all future tall buildings. CFD will be used to analyse how smoke and flames might spread through a

BELOW An infinity pool is a major attraction of Singapore's Marina Bay Sands sky garden. *(Sarah Ackerman)*

proposed design, allowing it to be tweaked to minimise any risk to life. Fire-resistant elevators and sky bridges could also be specified.

Although there is likely to be public access to sky courts and the like, this will not be unrestricted access. Any landmark skyscraper could be a terrorist target. Security will start with barriers and diversions that would prevent a truck bomb from being detonated close to the structure. Lessons learned during the IRA bombing campaigns of the past have been applied to the design of curtain walls. Sadly, the most iconic buildings will have to have airport-level security at the entrances – metal detectors, X-ray machines, security guards. Such is the world we now live in.

On a different level, occupants can expect to have individual control of heating and ventilation. Even opening windows are an option with double-skin curtain walling (*see* Keeping the weather out, p.52). However tall the tower is, however high the floor, there will be no unsettling sway. Residents could have privileged access to private sky courts and gardens, and dedicated elevators to whisk them up to the residential floors.

Currently, penthouse apartments in supertall towers in major cities are selling for eye-watering amounts of money. City centre land is expensive and supertall skyscrapers are incredibly costly to build. Developers have to recoup their investment from somewhere, and history shows that this is never easy; in fact it is not always a possible outcome. So will supertall towers become the 21st-century successor to the gated luxury estates that currently sequester the super rich?

It's all in the name

A recent boom in the number of skyscrapers built in the UK's capital city has been accompanied by an increasingly popular global eccentricity. From China to the UK a handful of notable skyscrapers, like it or not, have become more familiar by their ascribed nickname.

In Beijing the headquarters of China Central Television is housed in a particularly unorthodox building. With two leaning towers that are joined at the base and at the top the

BLOWING HOT AND COLD

Maintaining a comfortable year-round environment within a tall building is a well-developed technology. Modern heating, ventilating and air conditioning systems are available off the shelf from many suppliers. The necessary plant and equipment can be hidden away in basements and on rooftops, although some architects prefer to shroud rooftop plant and equipment behind decorative screening rather than leave it naked and unashamedly visible from any nearby taller buildings.

Internally, the ducts and pipework that distribute warm/cool air or hot/cold water can easily be concealed above suspended ceilings or below raised-access flooring. This approach works well with traditional skyscrapers under 30 storeys tall – but problems raise their heads when much taller towers are planned.

A single rooftop chiller unit would have to be enormous, for example. Heating plant in the basement would struggle to deliver warm air 100-plus storeys above. The solution that has evolved over the decades is to split the building into zones, typically at least ten floors high. Between each zone is a plant or mechanical floor, containing the essential plant and equipment to service each zone. Dedicated service elevators provide access.

These floors are often used to conceal structural bracing or tuned mass dampers. They also solve one knotty problem with very tall skyscrapers.

Every occupied floor, however high, needs potable water on tap. Residential floors also require supplies of hot water for bathrooms and kitchens. Automatic sprinkler systems must be permanently linked to an adequate supply of water in case of an outbreak of fire.

Basement-level water pumps would have to be massive to get water up to the highest floors – so the mechanical floors are used instead.

A modest-sized basement pump only has to supply a tank on the lowest mechanical floor. A second pump there sends water from the tank up to a tank on the next mechanical floor, and so on, forming a pump relay that ensures all floors receive equal levels of service.

With the move towards cities in the sky, these mechanical floors form a convenient transition between different building functions – retail, office, hotel, residential – where different levels of environmental control and fire protection would apply.

Mechanical floors can also solve one other less publicised problem. Some potential residents or commercial tenants would have a superstitious reluctance to live or work on the 13th floor. So the 13th floor becomes a mechanical floor with no public access – problem solved!

ABOVE China Central Television's headquarters is in an unorthodox building sometimes likened to a pair of 'big boxer shorts'. *(Shutterstock)*

OPPOSITE The Gherkin in London is thought to have been the trendsetter for skyscraper nicknaming. *(Shutterstock)*

shape has been given the somewhat tongue-in-cheek 'big boxer shorts' moniker.

But it is London (perhaps offering proof of its citizens' famously quirky sense of humour) that has adopted the sometimes flattering, sometimes not, idiosyncratic naming convention.

One of its most famous landmarks offers an early illustration of this. Big Ben has come to be known as the tower that offers a backdrop to countless tourist photos but in fact it is the bell, rather than the Elizabeth Tower that houses it, that is formally called Big Ben.

But the modern nicknaming trend is thought to have started with the 2004 arrival of 30 St Mary Axe, which is almost never referred to by its address but rather 'the Gherkin', thanks to its obvious cucumber-like form. Since then, London skyscraper nicknames have become almost *de rigueur*.

While it is pleasant to consider that sense of humour is the only reason for the emergence of ever more colourful names, there is a commercial gain to be made. Being afforded a catchy name has been a marketing gift for the office developer keen to stand out from the crowd.

The tallest of the city's buildings, The Shard, near London Bridge, has long been known by its seemingly informal title. Its design is

LEFT **The Shard was one of the first skyscrapers to opt for an informal-sounding name.** *(Shutterstock)*

OPPOSITE TOP **The 'Cheesegrater' (left), 'Scalpel' (second left) and 'Walkie Talkie' (right) have sprung up on the London skyline in recent years.** *(Shutterstock)*

OPPOSITE BOTTOM LEFT **The Sky Garden – a well-promoted feature of the 'Walkie Talkie' building.** *(Shutterstock)*

OPPOSITE BOTTOM RIGHT **While informally known as the less-than-glamorous 'Can of Ham', 70 St Mary Axe's owners have been less than keen to adopt the nickname.** *(Shutterstock)*

purposefully intended to create the sense that shards of glass are piercing the sky.

Another London inhabitant, at 52 Lime Street, followed The Shard's example and embraced its given nickname, 'the Scalpel', early on.

The 'Walkie Talkie' at 20 Fenchurch Street takes pride of place on the north-bank skyline of the River Thames, standing apart from its neighbours and affording viewers to clearly identify its outline.

Whether the familiar label has been fully embraced by the building's promoters is less clear as its marketing material references the Sky Garden, thanks to its unusual publicly accessible green roof space with restaurants and viewing platforms to entice visitors.

Some are even less delighted with their ascribed nicknames and anecdotally there are many who would like to forget the less-than-glamorous-sounding 'Cheesegrater' and 'Can of Ham' monikers. The former distinctly refers to itself in all official material as The Leadenhall on account of its 122 Leadenhall address and the latter is simply 70 St Mary Axe for the same reason.

A recently failed attempt to get planning permission for one of the capital's most unusual slender towers made no bones about trying to evoke a sense of charm about its odd, top-heavy design. But naming it 'the Tulip' was not enough to get it over the hurdles stipulated by the city's planners and so far plans for the 305m (1,000ft) tower have got nowhere.

Super, mega and quirky – the battle to stand out from the crowd

As skyscrapers became a more and more familiar sight, building designers have stretched the bounds of what is possible in an attempt to beat the competition. The result has been some of the world's tallest buildings – and some of the strangest.

At 318.9m (1,046ft) New York's iconic Art Deco Chrysler Building is the earliest example of a supertall skyscraper. At the time of its completion in 1930 it beat competition from the Bank of Manhattan Building – also finished that year – to gain the title of the world's tallest building. Perhaps more notoriously, having seen off one threat, the Chrysler was only able to hold on to its status for less than a year before the equally iconic 321m (1,250ft) Empire State Building stole the crown (*see* Empire State Building, 1931, p.126).

The latter managed rather better than its predecessor and carried the title for 40 years before the first of the World Trade Center's Twin Towers rose up.

And, of course, the devastating events of 9/11, when the world witnessed the downfall of these towers, left many believing the end of building tall was nigh. For some time there was a hiatus, but fortunately for tall-building enthusiasts, places like Dubai were soon to find themselves in the midst of an economic boom and there was enough money to build confidence that it was worth the risk of finding out whether developers and inhabitants wanted to live and work in supertall or megatall towers.

In 2010 the Burj Khalifa provided a defiant answer, opening as the world's tallest building, and at 828m (2,717ft) tall, surpassing its immediate predecessor, Taipei 101, by 320m (1,050ft) (*see* Burj Khalifa, 2010, p.149).

OPPOSITE New York's Chrysler Building and Empire State Building were the earliest examples of supertall skyscrapers. *(Shutterstock)*

BELOW The first of the World Trade Center's Twin Towers took the world's tallest skyscraper crown away from the Empire State Building after 40 years. *(Shutterstock)*

RIGHT AND FAR RIGHT Dubai's Burj Khalifa tower (left) decisively took the title away from the Taipei 101 building in Taiwan and, as the first to exceed 600m (1,969ft), became the trailblazer for a new category – the megatall skyscraper. *(Shutterstock)*

BELOW The Kingdom, now Jeddah, Tower set its sights on being the first mile-high building but soon revised its ambitions down to hitting the 1km (3,281ft) level. It is still under construction but work had stalled at the time of writing. *(Shutterstock)*

As is always the way, when one world record is set, in come others eyeing up ways to beat it. And so, well before the opening of the Burj Khalifa, the Mile-High Tower in Jeddah, Saudi Arabia, was already in development.

The name changed first to the Kingdom Tower and then to the Jeddah Tower as the height was deemed too economically challenging – contractors advised that although feasible, it was impractical. The challenge of mitigating the complex geology of the site and transporting materials to extreme heights led to the decision to shave around 500m (1,640ft) off the tower's height ambitions.

The Jeddah Tower design relies on a similar Y-shaped, triangular footprint and buttressed core to the Burj Khalifa, with its now tried-and-tested ability to create stability. But the Jeddah Tower tapers, rather than stepping back, as it rises. Construction was well under way on this new monolith, but at the time of writing (January 2020) had stalled and there is little clarity on when and how it will restart its climb to earn the accolade of being the world's tallest building.

Height is not always the factor that causes a skyscraper to hit the headlines. There is an abundance of architects and structural engineers who stay within the proven height boundaries, yet manage to express their creativity – and grab attention – through the shape and features of their buildings.

Although the concept for a moving, rotating skyscraper has been suggested more than once, a real working version has yet to emerge. But Sweden's Turning Torso in Malmö is evidence that creating the feel of a twisting, turning building is possible. Mostly residential units inhabit the 190m (623ft)-tall 'rotated' form. This visual quirk is achieved with nine five-storey-high cubes, joined and rotated a few degrees away from the block above and below. A circular internal spinal core featuring concrete cantilever slabs creates the stiffness and support, while a steel truss exoskeleton curves around the building's twisted form.

Another skyscraper that refuses to conform to the straight up and down idea of building tall is the Capital Gate building in Abu Dhabi. Fondly known as the 'Leaning Tower of Abu Dhabi', in a nod to that historic icon of Pisa,

The visual quirkiness of Sweden's Turning Torso, Malmö. *(Shutterstock)*

RIGHT The Leaning Tower of Pisa is now inviting comparisons to the Capital Gate building. *(Shutterstock)*

this 160m (520ft)-tall form appears to lurch upwards and out of the ground.

The building's 18° off-centre lean relies on two diagrids – diagonal load-bearing steel grids – to provide support as well as the careful design of the varied size and shape of the concrete floor slabs and to resist wind loads. While again the height is relatively modest, the building's astonishing lurching quality has earned it a place in the Guinness World Records in 2010 for being the 'farthest-leaning manmade structure'.

The leaning form is also a feature of the China Central Television Headquarters in Beijing. Unflatteringly inheriting the nickname of 'big boxer shorts' (*see* It's all in the name, p.55), the 234m (768ft)-tall building refuses to conform to expectations and, instead of a monolithic sky-high reach, aims to dazzle with its continuous looped form – a variation on the Möbius strip.

Two leaning towers are joined at the base and top with a 14-storey cantilever some 200m (656ft) above ground creating an astonishing overhang. The shape is unique and hits the mark when it comes to making a tower instantly recognisable.

Dubai's Burj Al Arab is an unquestionable success for the same reason, in the way its design evokes the form of a billowing

BELOW Abu Dhabi's Capital Gate building has an 18° off-centre lean. *(Shutterstock)*

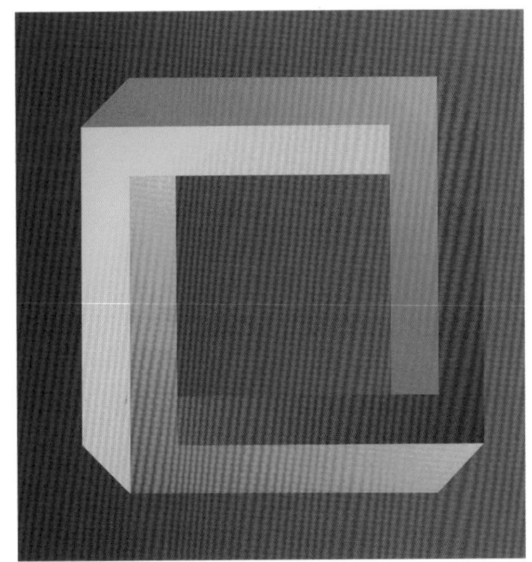

FAR LEFT **This Beijing building also features leaning towers, but in a continuous form, reminiscent of ...** *(Shutterstock)*

LEFT **... the Möbius strip.** *(Shutterstock)*

spinnaker sail of a yacht. The 321m (1,053ft), 56-storey uniquely curved form, with the addition of a jutting-out helipad, has become one of the most famous buildings in the world – a tricky challenge given it emerged amid the Dubai building boom. It is also an unquestionable engineering first with the inclusion of an 182m (597ft) high atrium that creates a dazzling entryway to one of the world's most opulent hotels.

And there are not many hotels that can boast having built a new island in the sea on which the hotel and its exclusive beach could be created.

FAR LEFT **Dubai's Burj Al Arab's design clearly evokes the billowing sail ...** *(Shutterstock)*

LEFT **... of a yacht.** *(Shutterstock)*

THE TALL STORY OF THE MILE-HIGH TOWER

The idea of being able to create a building that reaches a mile into the sky captures the imagination. Someone who was among the first to understand the allure of such a concept was renowned US architect Frank Lloyd Wright. His 1956 design for a mile-high building, known as 'The Illinois', was perhaps one of the most coherent early designs that attempted to make people believe such a feat might just be possible.

Of course it was never built, although architectural critics have noted a striking resemblance between that and today's tallest building, the Burj Khalifa. Lloyd Wright's 1,609m (5,280ft) skyscraper in Chicago was to have 15,000 parking spaces and 528 storeys accommodating 100,000 inhabitants. All would be served by a staggering 76 elevators.

Attempting to allay the inevitable criticism that it simply could never be engineered to safely stand so tall, Lloyd Wright claimed his elegantly tapered design, which emerged upward from a triangular base, would indeed be capable of withstanding the enormous wind loads that today's skyscraper engineers now well understand. The Burj Khalifa's Y-shaped base with a spiralling and tapering tower design suggests Lloyd Wright was likely on the right track.

The backdrop to Lloyd Wright's mid-20th-century idea was key. A design for a dizzyingly high skyscraper from a famous architect known for his love of rural life and a far more conflicted view of the modern city, would grab headlines. And a big reveal to a press gathering of this supercharged version of vertical living gave credence to the plan. It is also worth noting that the world record holder for tallest building at the time was New York's Empire State Building, which by then had been open for 25 years and reached a mere quarter of the proposed mile-high's vertiginous height at just 381m (1,250ft). Perhaps the scheme was too mind-bogglingly ambitious and certainly it was too far ahead of engineering innovations that have since made skyscrapers more cost-effective and less risky. Nevertheless, its boldness and vision for city living is one that today seems less fantastical.

LEFT Frank Lloyd Wright. *(Public domain)*

OPPOSITE Lloyd Wright unveils a 6.7m (22ft)-tall drawing of the mile-high building he proposed in 1956. *(Alamy)*

Chapter three

Anatomy of a modern skyscraper

Here we learn how skyscrapers are built and maintained, and how people are moved up and down them – and occasionally from side to side. Despite some tragic incidents since 9/11, the safety record of very tall buildings is actually very good – we consider here how designers can make them even safer.

OPPOSITE Steel reinforcing bars give strength to concrete pile foundations. *(Shutterstock)*

The need for good foundations

The eye-catching designs of skyscrapers around the world draw admiring gazes from passers-by and many a tourist attempting to take the finest skyline snapshot. But the towering simplicity of that which is visible above ground belies what is happening below ground to keep the structure stable.

Just like that portion of a tree above ground or the apex of an iceberg that can be seen above the water level, there is a large part that is hidden. A skyscraper's unseen foundations are the secret to how it stands so tall.

While it is generally true that low-rise buildings need only shallow footings for support, a skyscraper's loads are concentrated so much around its base that as much engineering is needed in the ground.

The heaviest-loaded parts of any skyscraper are the foundations. They not only stop the building from catastrophically falling over but also ensure support is provided to the entire structure in a way that creates minimal movement. This allows the building to function correctly and prevents serious impairments such as cracking walls or sloping door jambs. The foundations are where the building begins to take shape.

New York and Chicago, metaphorically at least, offered fertile ground for the earliest skyscrapers. But the types of soil and how they helped or hindered the build varied widely, generating the need for foundation engineering to develop rapidly to keep pace with demand.

Caisson foundations were the go-to foundation design for the majority of tall buildings from the boom of the late 19th century until the mid-20th century when the rise of the machines triggered the next evolution of foundations.

Early mechanical interventions involved a hybrid of hand digging efforts with some of the back-breaking work taken on by newly developed rigs. However, as with many

LEFT Like a tree's roots, a skyscraper's foundations are the hidden reason why these buildings can reach great heights. *(Shutterstock)*

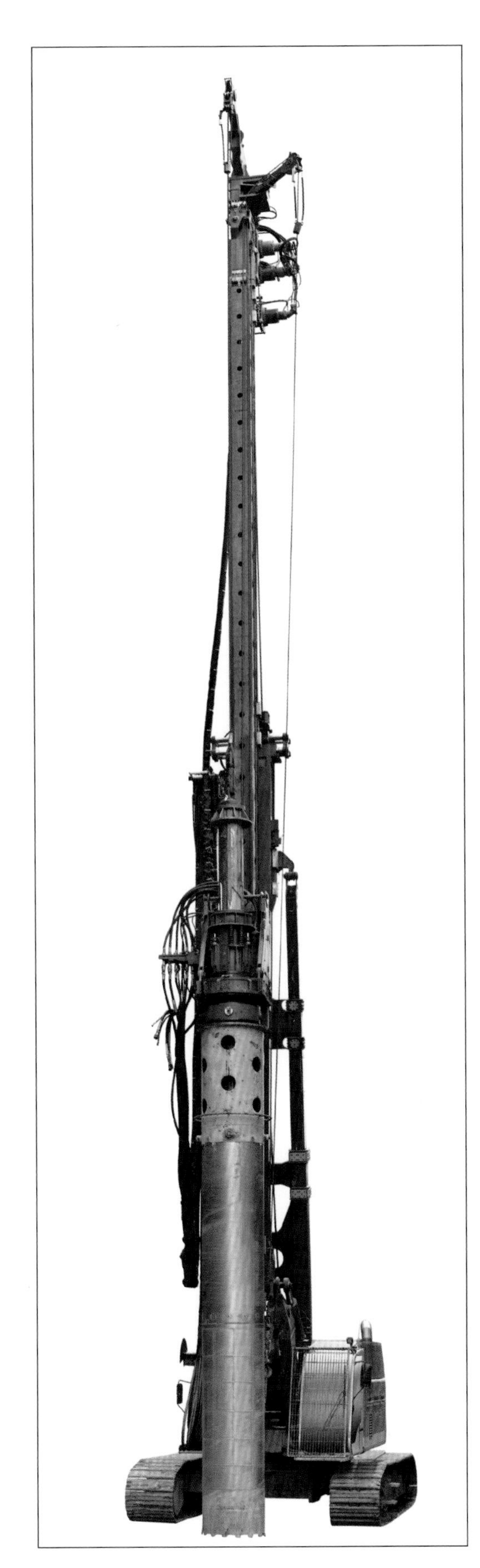

technological developments that thrived in the second half of the 20th century, piledriving rigs have since made building to ever greater heights more and more achievable.

Today's machines have incredible turning and driving power, enabling piling through the most complicated, hard and waterlogged ground conditions to great depths. High-tech instrumentation makes the job of pile-driving as far from gruelling as it was possible to imagine 100 years ago. In fact it is the method of installation that is the most important factor in determining the form a pile takes.

Most modern high-rise buildings sit on top of rotary-bored concrete pile foundations with steel inside to make them stronger. A broader

ABOVE A finished concrete pile with steel reinforcement bars protruding from its top. *(Shutterstock)*

LEFT A modern-day piling rig. *(Shutterstock)*

BELOW The range of cutting tools now available makes it possible to create large piles in a variety of soils and rocks. *(Shutterstock)*

term used to describe them is replacement piles, because the concrete poured into the ground to create them replaces the soil taken out. In fact, they are incredibly similar to their earliest caisson foundation cousins and their popularity today lies in their adaptability to be formed in the widest variety of soil and rock types and at extreme depths.

These specialist rigs are like huge drills and make it possible to form large-diameter piles at great depths. Where the ground is loose or waterlogged, steel casing can be rotated into the shaft to temporarily keep the sides from collapsing.

When the drill (or auger) reaches the required depth, known as the founding level, the shaft is cleared of soil with a special bucket attachment.

ABOVE Steel casing temporarily keeps loose soils from collapsing into the borehole and can be left in place to bolster the final strength of a pile. *(Shutterstock)*

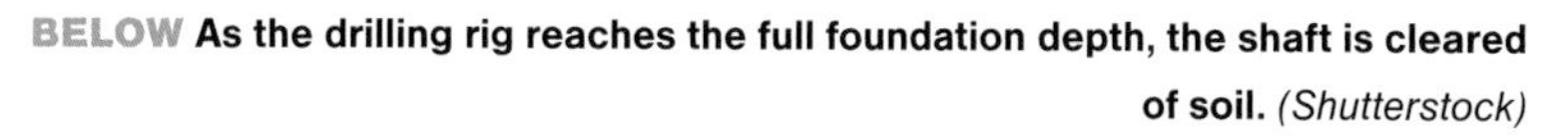

BELOW As the drilling rig reaches the full foundation depth, the shaft is cleared of soil. *(Shutterstock)*

A steel cage formed of reinforcing bars is lowered into the shaft to give additional load-bearing strength and concrete is poured into the base to fill up the hole and form the pile.

Before the concrete hardens, the temporary casing is removed, unless even greater support is needed, perhaps where soils are very weak, in which case the long steel tubes surrounding the concrete may be left in place.

Where ground conditions are made up of stiff clays and weak rocks, different cutting tools allow the shaft to be drilled out wider at the base, so-called under-reaming, before concreting to increase the load capacity at the lower part of the pile. A hinged tool attached to the bottom of the drilling rod (known as the Kelly bar) is expanded widthways into the ground when it reaches the full depth to form a bell bottom as it turns.

BELOW Steel reinforcement is lowered into a cleared-out shaft and concrete is poured in and around it to form a pile. *(Shutterstock)*

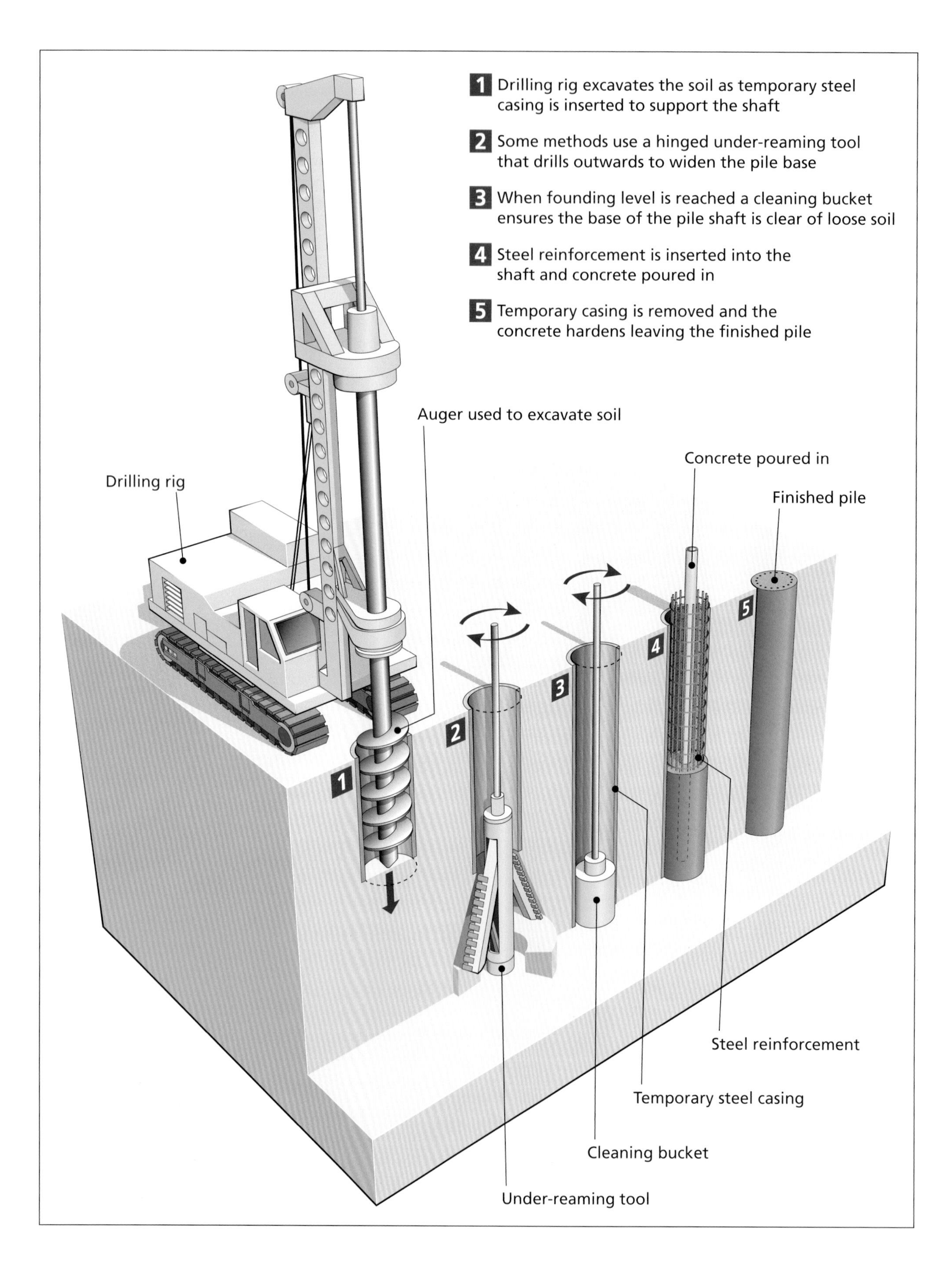
1 Drilling rig excavates the soil as temporary steel casing is inserted to support the shaft
2 Some methods use a hinged under-reaming tool that drills outwards to widen the pile base
3 When founding level is reached a cleaning bucket ensures the base of the pile shaft is clear of loose soil
4 Steel reinforcement is inserted into the shaft and concrete poured in
5 Temporary casing is removed and the concrete hardens leaving the finished pile
Auger used to excavate soil
Drilling rig
Concrete poured in
Finished pile
1
2
3
4
5
Steel reinforcement
Temporary steel casing
Cleaning bucket
Under-reaming tool

Improved understanding of the physics of building taller and taller means modern piles are capable of resisting loads from more than one direction.

End-bearing piles take vertical loads down to their base where they connect with hard rock. While this may seem to be the obvious type of pile, many, often in clay soils, rely on friction for support. The easy way to imagine how this works is to think of getting your foot stuck in mud and when you go to tug it out, the mud pulls your shoe off.

Loads from a building above are taken down through the friction piles and transferred out through their sides, rather than their ends, into the surrounding soils and rock. This works because there is enough tension between the sides of the pile and the surrounding earth that it is effectively being gripped by the soil to prevent it moving.

In fact, often the two types are combined and most of the vertical load on the longest piles typically needed for skyscrapers is supported by this skin friction – while the bottom of the pile carries only about a third of the total load.

continued on page 78

OPPOSITE Rotary-bored piles and how they are installed. *(Anthea Carter)*

LEFT The type of soil and rock beneath a new skyscraper determines the type of pile that is needed. *(Shutterstock)*

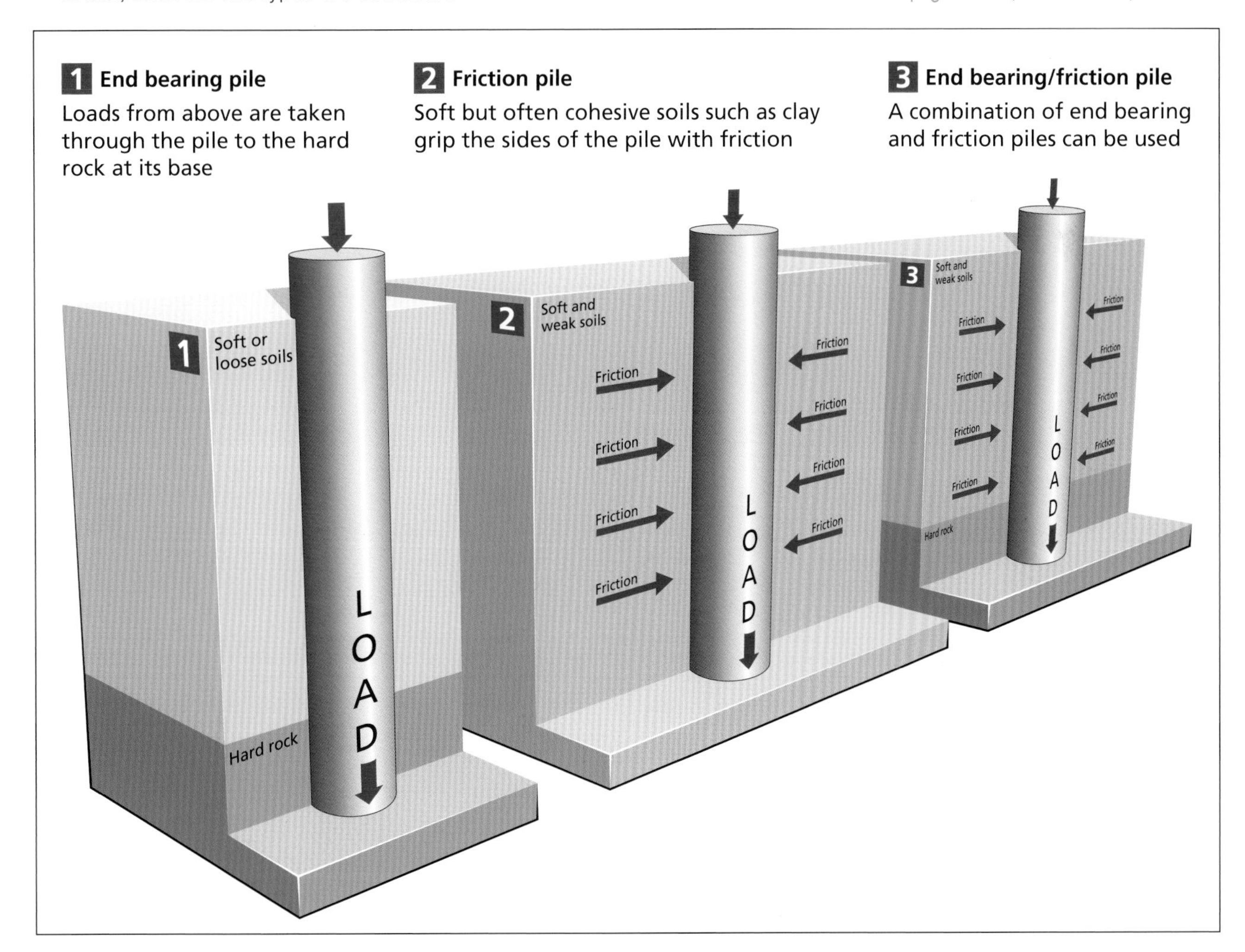

BELOW Piled foundations for skyscrapers allow for loads to be taken down to the bedrock, out to the sides, or sometimes a combination of both. *(Anthea Carter)*

Advances in foundation engineering make it feasible for today's skyscrapers to have multi-level basements. These below-ground structures – known as substructures – are increasingly cavernous spaces used for car parking, storage of mechanical and electrical equipment, gyms and perhaps even spas for high-flying hotel guests.

Most soil types make it easy enough to build a single-storey basement by simply digging out around the building's footprint and constructing the basement within. However, when digging deeper for more than two levels, the lateral loads from the soil and groundwater pressure will risk the excavation collapsing-in on itself. To resist this, engineers rely on two methods to build deeply and safely.

In some soil types and depths it is possible to use a system that forms walls of interconnecting piles around the substructure. Secant piled retaining walls are so called because of the way the piles intersect one another to form a continuous wall that retains soils and groundwater behind.

With a clear site and before excavation begins, piles are intermittently bored around the perimeter. Once the concrete is in place for the first piles, then a second round of piling is performed between the gaps, with the drilling auger cutting an arc into the concrete of the earlier installed piles. Often, one intersecting set is made less robust than the other, either because it uses a lower strength concrete or is free of reinforcement. Some of the piles need to be very robust to take the loads from the skyscraper above but the rest simply form a cut-off from the retained soil and water. Consequently, secant piles are described as combining soft, firm and hard piles, depending on the varying robustness of the steel reinforcement in the pile.

With the soil effectively held in place behind the secant walls, excavations can take place within to allow construction of the basement floors to follow on. The UK's tallest building, The Shard, completed in 2013, offers one recent example of this kind of basement construction. Its 95-storey above-ground

BELOW Secant piled walls are formed from sequential rows of piles that intersect one another and retain soil and groundwater. *(Shutterstock)*

superstructure relies on 388 secant piles with a diameter of around 900mm (34in) to form a three-level basement and provide stability.

Sometimes piling is not the most practical answer for deep basements. The machinery can disturb nearby buildings and the process requires a degree of easy access not always readily available for the vast equipment needed to dig deep. In that case, and particularly in the case of substructures that reach depths of more than 40m (131ft), these large basement builds use the diaphragm walling method.

Diaphragm walls are suitable for most subsoils and their installation generates minimal vibration and noise. While similar in their function to the secant piled wall, diaphragm walls use different equipment and are formed from in situ concrete.

Crane-mounted grabbing tools take large 'bites' from the soil around the basement outline to form holes in the ground.

These holes are kept stable with a heavier-than-water 'bentonite' clay slurry to enable the final wall depths to be reached. Concreting and steel reinforcement form the walls within these dug-out sections. The gel-like bentonite slurry is displaced as the concrete is poured and is usually filtered and recycled.

With both diaphragm and secant walls, once the reinforced concrete is in place and ready, construction of the basement floor slabs continues either via a bottom-up or top-down sequence.

The former relies on vast hydraulic bracing installed at intervals as excavations advance downward. These work like large props that push the walls out and hold back the soil and groundwater and can be adjusted with hydraulics to replace the pressure previously supplied by the soil that has been removed.

Top-down construction, by contrast, does away with props and instead casts the basement floor slabs in reverse order, from the upper basement down to the lowest level. This way, completed floor slabs stabilise the walls, while excavation continues below via holes in the floor slab. These spaces within the floors will often be repurposed as stair wells or elevator shafts servicing the future inhabitants of the building.

LEFT Crane-mounted tools take 'bites' out of the ground to form diaphragm walls. *(Shutterstock)*

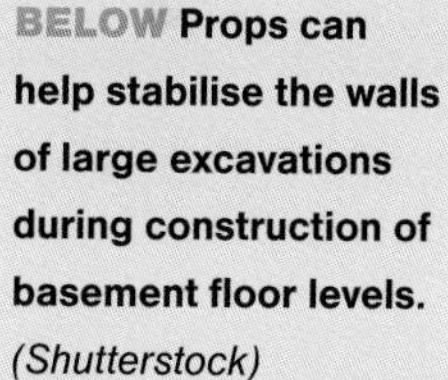

BELOW Props can help stabilise the walls of large excavations during construction of basement floor levels. *(Shutterstock)*

ABOVE Soils are extracted from the ground and analysed before foundation design takes place. *(Shutterstock)*

The taller the structure, the more important it is for today's foundation engineers to design for uplift and lateral loads. Uplift can be caused by the pressure of groundwater wanting to push the foundations upwards and out of the ground, but also from the force of wind wanting to push the above-ground tower and its foundations over. Both would cause any skyscraper to lean or even topple over if not designed to counteract these forces. In this case, often when the soil is made of mostly clay, under-reamed piles help to anchor the piles, and the skyscraper, against the uplift.

Great engineering effort is made to understand the soil and rock formations that lie below a skyscraper through early and comprehensive site investigation work. Boreholes are dug deep and soil samples extracted and analysed across the site to give engineers as full an understanding as possible. Then, working alongside the structural engineers and building designers, they can draw on all their expertise to define the technique, diameter, depth, steel reinforcement and number of piles needed for every skyscraper.

THE WORLD'S BIGGEST 'BATHTUB'

The extraordinary and enduring story of the original World Trade Center's Twin Towers in New York began long ago with the design and construction of its unusual substructure and foundations.

The basement of the original World Trade Center covered around two-thirds of the 159m by 299m (520ft by 980ft) site and it had to perform a wide variety of functions. Firstly, it had to contend with keeping out water from the nearby Hudson River. This runs along the west flank of what was the 6.5ha (16 acre) site and caused much of the ground beneath the area to be saturated, even at just a few feet below ground level. It is not without irony that the vast substructure would come to be known as 'The Bathtub', given that it was actually designed to keep the water out rather than to allow it to fill up.

Secondly, the basement was also needed to provide for the foundations for the two iconic 110-storey towers. These would stretch down to the Manhattan schist bedrock at depths ranging from 16.8m–24m (55ft–80ft) below the surface. Finally, the seven-level substructure would also accommodate parking and a train station and provide protection for existing subway and metro tunnels.

The Port Authority, which owned the site, put its faith in a brand-new technique not widely used outside underground metro rail projects during the late 1960s construction era. Here, however, it was to be used on a grand scale not yet seen in building construction to provide a watertight basement for the whole development.

A slurry (also called diaphragm) wall, with a total perimeter length of 1,067m (3,500ft), was the answer. Site workers dug 900mm by 6.7m (3ft by 22ft) slots down to the bedrock, filling the voids with bentonite clay slurry as they advanced. This liquid, being slightly heavier than water, provided temporary support for each dug-out slot, preventing the soil

LEFT The former World Trade Center Twin Towers had an unusual substructure. *(Shutterstock)*

caving in and keeping the groundwater out. Once at the required depth, vast steel reinforcing cages (each weighing up to 20t/22 US tons) were lowered in. The wall panels were formed via a 'Tremie' pipe that carried the wet concrete down through the bentonite to the base of the slot. As concrete poured in, the bentonite would be pushed up and pumped away.

All 158 adjoining panels were built in the same way over a single year and the next year was spent digging out the more than 765,000m^3 (1,000,000yd^3) of soil from within the slurry wall structure. However, the vast substructure needed lateral support in between its walls to prevent them collapsing as the soil was dug out. For this the team came up with a plan to insert a vast array of 1,500 high-strength 'tendon' anchors to 'tie back' the walls.

These were threaded through steel pipe casings drilled into the walls and down at sloping angles to 10.7m (35ft) into the bedrock. Once in place, the tendons could be tensioned to help create anchoring resistance before being grouted in. Each was able to take loads of up to 272t (300 US tons) during the excavation process.

Once all the soil had been removed, the basement floor levels were constructed, providing permanent lateral strength and

BELOW A vast slurry wall created foundations for the towers down to the bedrock and kept the site watertight. *(Eric Ascalon)*

RIGHT A segment of the original 'bathtub' wall was restored and remains on display at the 9/11 Memorial & Museum in New York. *(Shutterstock)*

enabling the tension in the anchors to be released. The casings were sealed over as they had performed their duty and were no longer needed.

The walls remained robust for decades. Incredibly they survived intact in 1993 when terrorists set off a bomb in the basement near to a column supporting the North Tower (*see* Dealing with disasters, p.110). Despite some damage to the basement floors supporting them, the walls themselves stood firm and refused to let any water in. A visual check on their condition in the spring of 2001 found that they remained in good condition.

But just a few months after that inspection came the devastating 9/11 attack, when two hijacked planes crashed into one tower and then the next, triggering their catastrophic collapse. It was too much for the walls to contend with. Despite their resilience and remaining intact across much of the site, their structural integrity was compromised to such an extent that a new 'bathtub' had to be built to enable the ensuing redevelopment of the World Trade Center site.

However, there remains a sign of their resilience today. One segment of the original bathtub was restored and is on display at the 9/11 Memorial & Museum, which also inhabits the revived Manhattan landmark.

Standing tall – the structural frame

At the beginning of the skyscraper age, the engineers responsible for designing the all-important structural frame had no computers, no wind tunnels and no design codes. Working with slide rules alone, making up their own codes and standards as they went along, these first-generation structural engineers inevitably took a 'belt and braces' approach. Their buildings were over-designed by modern standards, but they succeeded in reassuring developers and occupants alike and paving the way for even taller towers.

Skyscraper design has evolved a long way since then. The pressure to build ever taller forced engineers to look towards radical solutions. What worked for a 20-storey building would be uneconomic or even unsafe for a tower in excess of 50 floors. As buildings get taller, wind loading begins to dominate the design. In particular, designers have to consider the comfort of the occupants.

Human beings generally accept that ships, cars, aeroplanes and trains shake, vibrate, lurch up and down, sway from side to side, accelerate and decelerate. Only in extreme cases will passengers panic or become nauseous, although human tolerance of motion and acceleration varies widely from individual to individual. Buildings summon very different feelings.

Subconsciously, occupants expect all buildings to be stable, static and safe. Although all buildings will move slightly in response to high winds, on low-rise structures these movements will be imperceptible. As buildings get taller and more slender, wind-induced sway can become obvious to the occupants, especially to those on the higher floors. At the top of a 100-storey building sway of up to 1m (3.3ft) or more each way could be expected, cycling back and forth every 10 seconds.

This is perfectly safe and will have been designed for. The oscillation is caused by wind vortices along the sides of the building, rather than wind pressure on the windward face. These can cause the building to resonate at its natural frequency, intensifying the degree of sway.

As the sway builds up, occupants of the higher floors will often first become aware of the movement from visual clues rather than any physical reaction. Light fittings may start to swing, wine in glasses might start to slosh about and the building could creak and rattle. This can cause genuine alarm and fears that the building is close to collapse.

At higher levels of sway nausea will affect a significant percentage of the occupants. Eventually it will become difficult to walk, or even stay upright. By this point debilitating fear and panic will have set in.

Structural engineers have always added some form of bracing to the structure that carries the vertical loads. This bracing, designed to resist both wind and earthquake lateral loads, started out as strengthened walls, known as shear walls. Hotels and residential towers have a cellular structure so reinforced partition walls could provide all the stiffening needed on medium-rise towers up to the 30-storey-mark. Typically these would be formed from reinforced concrete, although solid steel plate shear walls are now becoming popular in areas of high seismic activity, such as California.

Advantages claimed for them include speed of erection (particularly when the steel panels are prefabricated offsite), reduced weight and significant space saving over the much thicker concrete alternative. They are also becoming popular in colder regions, where concreting operations are inhibited by low temperatures.

Commercial office developers demand open office floors. Instead of obtrusive shear walls, designers typically opted to concentrate all the lateral resistance needed in a central reinforced concrete service core. This housed the elevators, the emergency stairs and usually the bathrooms, and supported the inner ends of floor beams whose outer ends would be supported by perimeter columns.

As buildings grew taller, however, this simple solution became increasingly impractical. The central core got larger and larger, not just to resist increased lateral loads, but also to house the many more elevators taller skyscrapers required. Various alternatives featuring internal steel bracing were tried, but none offered a satisfactory solution to heights above the 50-floor mark.

Innovative solutions to the elevator challenge were eventually developed (*see* People movers, p.101). Major structural innovation was also needed if skyscrapers were to continue to evolve. In the 1960s a structural engineer in Chicago, the birthplace of the skyscraper, came up with a completely radical approach to tall-building design (*see* Willis Tower, 1973, p.136). This was the concept of the vertical cantilever.

Cantilevers have been around for millennia. Basically a cantilever is a structural member anchored at one end only – like a diving board or flagpole. The most efficient structural form for resisting loading on a cantilever is the tube – think of bamboo or a wind turbine tower. Thus a skyscraper anchored to its foundations is a vertical cantilever, and the closer its design approaches some form of tube, the more efficient it will be.

A circular cross-section is not the only option. Square or rectangular floor plans are the most common; there have even been skyscrapers with very inconvenient triangular floor plans.

In essence, the common factor is that all the horizontal load resistance is now down to the perimeter structure, usually formed of closely spaced columns tied together by deep beams. Usually there are a small number of internal columns carrying vertical loads only at the core. This frees up a lot of internal floor space, but it does spell the end of the all-glass curtain wall, as only around half the exterior surface is available for windows.

A more open exterior can be achieved with a braced or trussed tube design. In this, the vertical columns are further apart but linked by diagonal bracing. This bracing can be made into an architectural feature, as on the 100-floor John Hancock Center (recently renamed 875 North Michigan Avenue) in Chicago, completed in 1969. For some, however, this style of bracing is too visually obtrusive, and a different solution must be sought.

A tube-in-tube design retains the external framing of the outer tube and adds an inner

RIGHT **Bracing as an architectural feature, seen here on Chicago's John Hancock Center, is not to everyone's taste.** *(Joe Ravi)*

LEFT Manhattan's 182m (597ft) Hearst Tower illustrates the potential of diagrids. *(OptimumPx)*

tube – which could be reinforced concrete or steel framed – that houses the elevators, but is designed to shoulder a significant proportion of the lateral loads. Normally, inner and outer tubes will be linked by a number of substantial 'outrigger trusses', which will connect with horizontal 'belt trusses' running around both inner and outer tubes. The levels at which the outrigger trusses are located are usually adopted as mechanical service floors with no public access.

An efficient cantilever tapers from its anchorage, where stresses are highest, out to a low-stressed tip. Virtually all the first wave of very tall 'tubular' skyscrapers was severely rectangular and monolithic. The building that took the tallest building crown in 1974 soon after opening in 1973 looked very different.

Dubbed the 'bundled tube', the Sears Tower (now called Willis Tower) was based on nine individual square tubes linked together. Only one went all the way to the uppermost 108th floor, the rest stopping off at different levels to create a close approximation to an efficient cantilever.

Another way of looking at the structure would be to consider it as a giant steel box beam with four webs and four flanges. Whatever the approach, the new concept gave architects much greater freedom, as the individual tubes could be arranged in a number of ways.

In fact the Willis Tower is about as tall as a bundled tube structure could go without compromising both its practicality and its economic viability. An increasingly popular alternative is the tube based on an external diagonal grid – or diagrid – of steel members. This diagrid resists both vertical and horizontal loads, and with the latest computer-aided design techniques diagrid tubes can be twisted or curved into quirky shapes.

Adding enough wind bracing to make the building very stiff and highly unlikely to sway significantly even in hurricane winds has its

RIGHT Diagrid construction and a smooth tapering profile combine in China's 439m (1,439ft) Guangzhou International Finance Centre. *(慕尼黑啤酒)*

drawbacks, however. It could compromise the architecture, be complex and expensive (progressively so as the number of storeys increases) and could reduce usable interior space to uneconomic levels.

Alternatively, designers could put a model of the proposed design into a wind tunnel, although these days it is claimed that more accurate predictions of how a skyscraper will behave in high winds can be achieved by CFD. The objective of both techniques is to shape the external façades of the building to prevent the build-up of vortices that could cause it to sway beyond what occupants can tolerate.

This is one of the inherent benefits of the bundled tube design, with the setback sections disrupting the dangerous vortices. Setbacks also feature on the record-breaking Burj Khalifa (*see* Burj Khalifa, 2010, p.149), but for the proposed 1km (3,281ft) Jeddah Tower in Saudi Arabia – formerly known as the Kingdom Tower – a smooth taper was selected.

Such a profile is the most effective way of smoothing out the airflow round a supertall skyscraper. Minimising sway is particularly important for the Jeddah Tower, as it is to be a mixed-use building with residential accommodation on the upper floors. A smooth, tapered profile is more complex and expensive to construct, but on such an extreme building it is almost essential.

Both the Burj Khalifa and the Jeddah Tower feature a new structural form, the buttressed core. On both buildings the floor plan is Y-shaped, creating internal spaces with increased window views and avoiding excessive floor areas.

Some of the designs that can successfully 'confuse' dangerous vortices can be difficult or expensive to build, or can seriously reduce the project's economic viability. A less intrusive approach is to add some form of sway damper internally, usually close to the highest floor, which can automatically restrict sway to tolerable levels even in typhoon conditions.

Panasonic

There are two basic types in use at the moment: the tuned mass damper (TMD) and the tuned liquid damper (TLD). Both are 'tuned' to the building's natural frequency. A TMD uses a massive pendulum weighing up to 1,000t (1,100 US tons), whose swing is restrained by springs and hydraulic dampers. By contrast, the lower-tech TLD is basically a tank of water fitted with internal baffles.

Both work on the same principle. As the building begins to sway, the pendulum starts to swing or the water begins to slosh backwards and forwards. The hydraulic dampers or the internal baffles result in the movement being out of phase with the building's natural harmonic, thus cancelling out virtually all the sway.

A TLD has the added benefit of potentially acting as a back-up reservoir for the firefighting sprinkler system. As unlikely as it may seem, the 660t (728 US ton), 5.5m (18ft)-diameter steel TMD suspended from the 92nd floor of the 508m (1,667ft)-tall Taipei 101 office skyscraper in Taiwan is a popular tourist attraction. It has its own mascot – Damper Baby – that has even spawned its own website and comic book.

Floors are part of the structural frame, unlike the cladding, which is almost invariably

OPPOSITE The world's tallest building from 2004 to 2010, Taipei 101 tower has to withstand frequent typhoons. *(Archive team)*

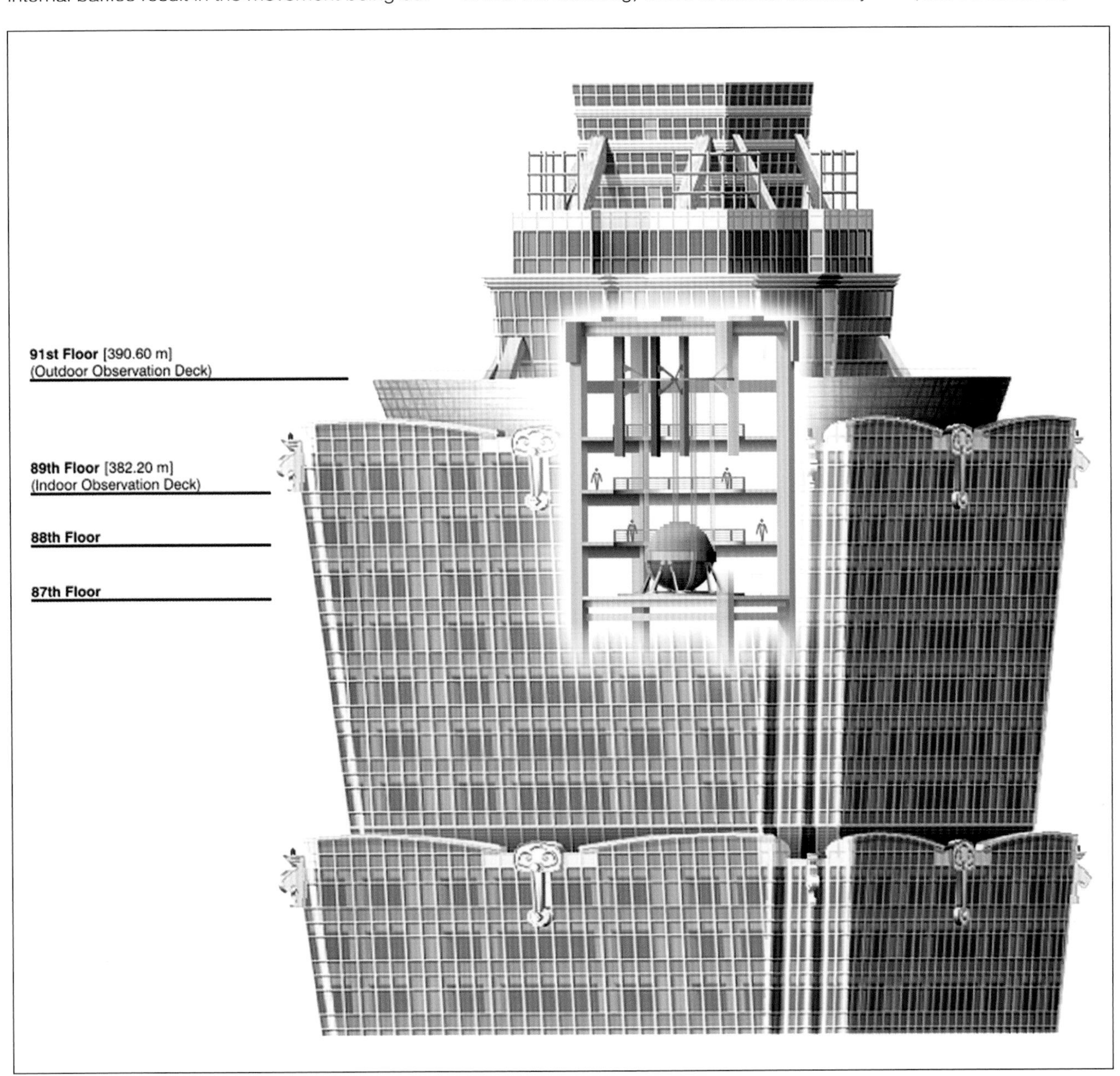

BELOW Taipei 101's stability depends on a tuned mass damper. *(Somekindofhuman)*

CONCRETE OR STEEL?

First-generation skyscrapers had primary frames of steel. Wind-bracing was achieved by the use of steel as well. Eventually, designers started using cheaper reinforced concrete shear walls to resist lateral loads. Back then the open plan office floor was literally inconceivable; instead, office blocks were divided up into a large number of offices of varying sizes, which created straightforward opportunities for inserting reinforced partitions to act as shear walls.

At the end of the 19th century, concrete technology was in its infancy. What we know as Portland cement had only become available in bulk in the 1890s; its properties were little understood and its performance was very variable. Concrete had to be mixed on site in small batches. It remained workable for a comparatively short time, around two hours on average, but this was much shorter in high ambient temperatures. In low temperatures it would set and gain strength very slowly, and below freezing point it would need special protection otherwise it would gain no strength at all.

Nevertheless, as skyscrapers continued to evolve, concrete began to take on a more significant role. First came concrete floors. Then designers began to see the advantages of concentrating the building's shear resistance into a reinforced concrete service core. As the technology continued to develop, concrete became stronger and more predictable, particularly so since the 1960s and 1970s.

Another development that helped promote increased use of concrete was the arrival of ready-mixed concrete. Ready-mixed concrete depots sprang up around major cities, with truck mixers delivering fresh concrete straight to the crane skips that would lift it into place on the project. No need to set up a site mixer with its cement silos and aggregate bays, so much more room on site. Generally, ready-mixed concrete was a better product than anything that could be mixed on site, thanks mainly to the producers' familiarity with the local cement and aggregates.

Ready-mixed concrete made large pours on pile caps a realistic solution. Using concrete as an alternative to steel for the primary structural frame took a long time to become routine.

Using concrete with the strengths available in the 20th century resulted in a structure

RIGHT Believe it or not, this massive tuned mass damper has its own mascot, website and comic book. *(Armand du Plessis)*

that would be significantly heavier than the steel alternative. Foundations would have to be much larger, making erection times considerably longer. Only when the choice was between expensive imported structural steel sections or locally produced ready-mixed concrete did the sums add up.

One problem of combining a steel primary structure with a reinforced concrete core as a design solution for a supertall skyscraper is that concrete, believe it or not, is not a truly stable material like steel. Under constant vertical loads the core walls will gradually shorten – known as 'creep'. Differential movement between the core and the steel perimeter columns can cause all sorts of problems. Designing this out is very challenging. Even today, creep is hard to predict.

Nevertheless, concrete is increasingly seen as a realistic option for the structural frame, particularly for perimeter columns. Advanced mixes offer much higher strengths than ever before, with much-improved consistency when freshly mixed. The development of very powerful concrete pumps capable of delivering workable concrete to 1km (3,281ft) above ground level speeded up construction significantly (*see* Planning the project, p.91). And concrete's inherent fire resistance is another plus.

Steel will always be lighter than concrete in many applications. Steel does not creep, and as the product of a highly sophisticated manufacturing process its performance in service is very predictable. Erection can be very rapid. Yes, its vulnerability in a fire situation is always going to remain, although there are many options for providing economical fire protection (*see* Dealing with disasters, p.110).

In a situation where steel and concrete are readily available, the best solution for a supertall megatall – or even a hypertall building – might be a hybrid frame. Core and columns would be high-strength concrete pumped into place, while floor beams and outrigger trusses would be steel.

Most of the world's tallest buildings are hybrids. The Willis Tower is the tallest all-steel skyscraper – and almost certainly the last record-breaker to go down that route. It is an all-concrete building that holds the record at the time of writing – the Burj Khalifa (*see* Burj Khalifa, 2010, p.149). Nevertheless, the hybrid solution is likely to dominate skyscraper design for the immediate future.

non-structural (*see* Life in the sky, p.49). Floors make the largest contribution to a multi-storey building's self-weight – known as the dead load – as opposed to the live load of occupants, furnishings, equipment, storage and so on. As skyscrapers through history grew ever taller and the number of storeys passed the 100-mark, much attention was paid to minimising the weight of the floors without compromising sound attenuation, fire resistance or occupant comfort. Lighter floors meant slimmer structural frames, in turn generating less load on the foundations – all of which resulted in lower costs.

Slimmer floors lead to reduced floor-to-floor heights. On a tall building whose maximum height is limited by planning constraints, these reductions could enable an extra floor or two to be squeezed in, enhancing the building's commercial potential.

Residential floors carry the lowest live load and are usually the slimmest. Some form of concrete slab is the norm, but there are several options. These are all designed to deal with concrete's Achilles heel – its weakness in tension.

A beam or a floor under load will sag slightly. The upper surface will be compressed, while tension will develop at the lower surface. A floor made of mass concrete will begin to crack in the tension zone, and eventually the cracks will lead to a disastrous collapse.

Traditionally, steel bars are cast into the concrete to carry the tension forces. This works – but simply reinforced concrete floors tend to be thick and heavy, especially over long spans. The most popular current alternative for residential skyscraper floors is post-tensioned, prestressed concrete.

Prestressing involves passing steel

cables – or tendons – through ducts cast into the concrete. Then, when the concrete has hardened sufficiently, the tendons are tensioned, thus compressing the concrete. If this prestressing is high enough, the tension forces in the concrete will never exceed the compressive forces induced by the prestressing, so the concrete never cracks. Slim prestressed concrete floors tend to be more flexible than other options, so deflections could be a problem in high-load areas. A stiffer, high-strength alternative is the composite slab.

Self-supporting profiled galvanised steel moulds up to 225mm (8.9in) deep are first installed, then concrete is poured into them. Despite the metal being only 2mm (0.08in) thick or less, the shape of the moulds makes them self-supporting over spans of as much as 6m (19.7ft), so no additional propping is needed during the concrete pour.

Once the concrete has hardened, the moulds provide the tensile reinforcement. Achieving a 120-minute fire resistance is possible without additional fire protection to the exposed steel. The system also works with

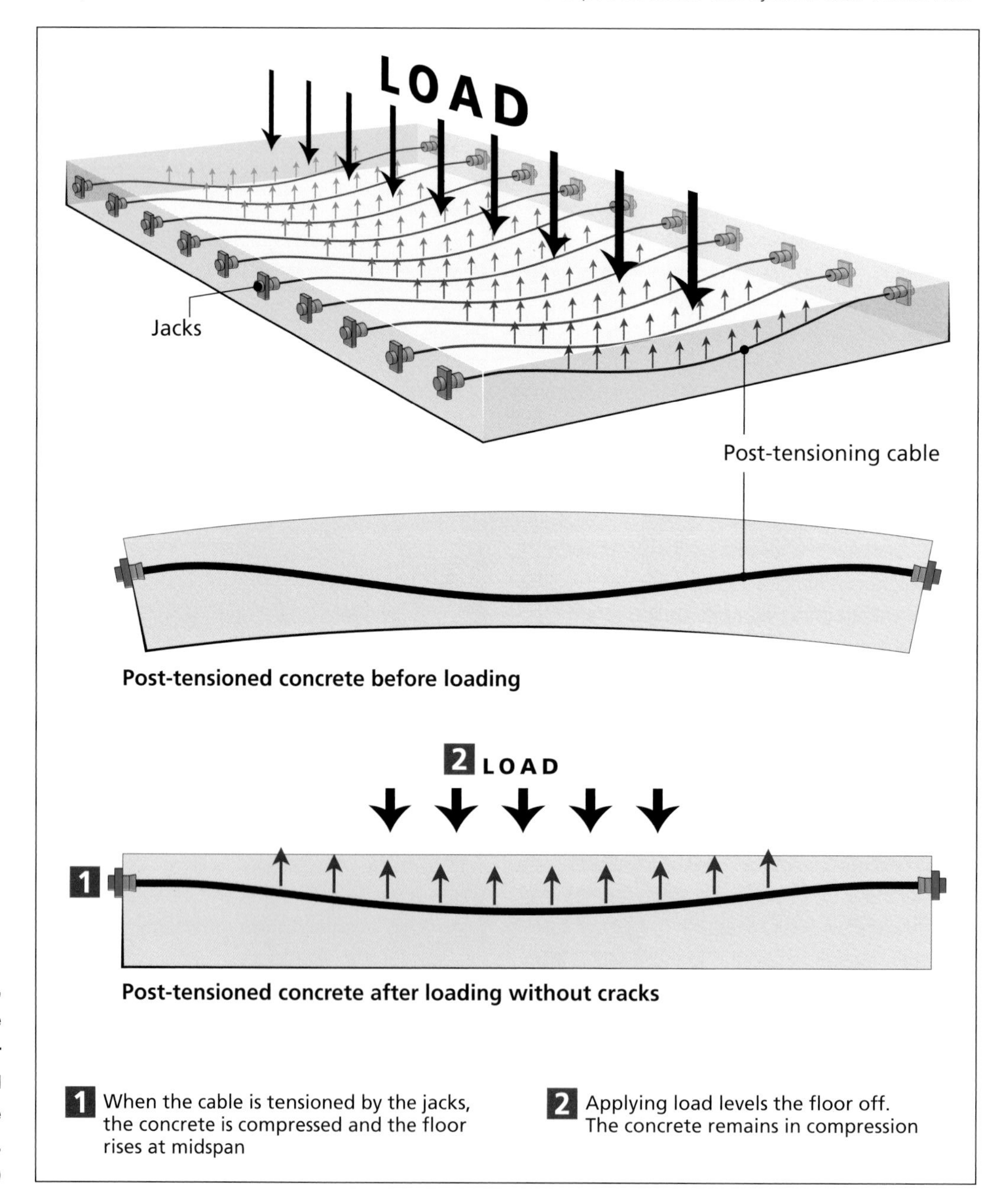

RIGHT Post-tensioned, restressed floors are slimmer and lighter than the conventional reinforced concrete alternative. *(Anthea Carter)*

lightweight concrete, where factory-produced expanded clay pellets replace the coarse stone aggregate in the concrete.

Structural floors in office areas are normally topped with raised access floors – basically a series of panels supported on metal frames. These facilitate the installation of all the cables and services the modern office requires.

Planning the project

Relatively few skyscrapers are constructed on greenfield sites. For the majority, the project begins with the demolition of existing buildings. In sensitive city-centre locations this can be a serious challenge. There will be issues with noise, safety, dust and traffic congestion. These days as much as possible of the demolition waste will be recycled in situ, but any contaminated soil will have to be removed and taken to an approved disposal facility. The discovery of asbestos in the existing buildings will also cause complication and delays.

A secure perimeter barrier has to be established as soon as possible, with controlled access. Services must be connected, and security lights and security cameras installed. There must be negotiations with the city authorities and the local police to agree restrictions on heavy vehicle movements, local road cleaning, permitted noise levels and site working hours. In historic city centres an archaeological investigation may be mandated as well, with potential major delays if significant finds are made.

From the outset, one of the biggest initial challenges is how to deal with all the masses of materials and components required for a typical skyscraper project. Most skyscrapers are constructed in city centres on very restricted sites accessible only by road. Space to store significant stockpiles of materials and components will be minimal. 'Just in time' deliveries may be possible for certain components on some projects, and the use of concrete mixed elsewhere and delivered to the site in truck mixers only when needed will reduce the logistical pressure – though it will not eliminate it.

One common solution is to rent a storage area on the fringes of the city, close to the main road network. Here, components and materials can be stored in bulk, ready to be collected and delivered to the site proper when called for. On the largest projects this storage area could even include its own concrete batching plant supported by a fleet of dedicated truck mixers.

Massive volumes of freshly mixed concrete

BELOW Component and materials handling were one of the biggest challenges on the Burj Khalifa project. *(Public domain)*

are usually needed for foundation and basement construction (*see* The need for good foundations, p.70). Organising a steady truck mixer shuttle service for very large pours will be another major test, given the potential traffic congestion on city streets and the limited 'shelf life' of freshly mixed concrete. However, it is when the foundations are complete and erection of the structural frame is ready to start that the real challenges begin.

Important decisions have to be made, especially critical on supertall projects. What type and number of cranes will be needed (*see* Sky-high lifting, p.93)? Does it need a complex concrete core? Should it be a jump form construction system, or would slip form be more effective (*see* Slipping up, p.96)? Is there the expertise and equipment locally available to provide reliable concrete pumping to all floors in the project (*see* Pump it up, p.100)?

Providing efficient means of access for all the hundreds, sometimes thousands, of workers who could be involved on a supertall project is just as important as ensuring materials and components are lifted into place smoothly and safely. A jump-lift system could be chosen, or reliance placed on temporary self-propelled climbing elevators mounted externally. As the building grows taller, the transit time for operatives at the top levels becomes unacceptably long. The only option is to set up 'comfort stations' within the structure as it grows, where can be found bathrooms, potable water, first-aid facilities, shelter from extreme weather and even hot food and drink.

Temporary offices for the engineers, construction managers and architects who need to be stationed close to the 'coal-face' could also be seen as essential. All such temporary habitations need to be serviced and have emergency evacuation routes.

As construction reaches ever higher, winds become stronger. Freshly placed concrete, typically in floors, is vulnerable to the drying effects of strong winds. As moisture is sucked out of the surface of fresh concrete, it shrinks and cracks, and without sufficient mixing water to complete the hydration process the surface concrete also loses strength. Exposure to strong sunlight can have similar effects, too. This being the case, temporary protection is needed, typically in the form of tented covers, which have to be strong and well anchored.

High ambient temperatures can accelerate the hardening of the concrete to the point that it becomes unworkable before the pour is finished – it could even harden and block the delivery lines of any concrete pumps in use. Using crushed ice in place of normal mixing water to batch the concrete is usually enough to control this problem. Conversely, in cold climates the concrete will take an unacceptably long time to develop strength, slowing the construction process significantly. There are now a number of heated formwork systems on the market that can insulate the fresh concrete from even sub-zero temperatures.

Traditionally, a topping-out ceremony is held when the basic structure is complete. On all-steel buildings, particularly in the USA, the last beam to be installed is painted white and signed by everyone on the project, while a fir tree is placed at the highest point. This practice harks back to ancient Scandinavian traditions. Again it was traditional for the developers of the project to install barrels of beer on the highest floor for all the site workers. These days these ceremonies are seen as ideal public relations events, particularly on landmark buildings, with celebrities and politicians invited to attend and enjoy lavish catering and entertainments.

Such ceremonies occur many months before the building is actually handed over to the client. Cladding and fitting out are complex projects in themselves, even more so on supertall mixed-use skyscrapers. Structural completion is just one milestone, albeit arguably the most significant one. The major challenges will have been overcome and the scale of the new skyscraper will be clear for the first time. For much of the workforce this will be the end of their involvement. When the building they have erected goes on to become a well-known landmark, possibly with an affectionate nickname, they will have something to be proud of for the rest of their lives.

SKY-HIGH LIFTING

As Dubai's Burj Khalifa approached its world record height of 828m (2,717ft), perched right at its highest level were two tower cranes, almost certainly the highest ever. Tower cranes are indispensable tools on every skyscraper project. It is even claimed that the state of a country's economy can be easily assessed simply by counting the number of busy tower cranes in the capital city. What is true is that these slender, elegant, yet simple machines are very efficient and economic movers of materials and equipment to great heights.

At first sight a typical free-standing tower crane may look impossibly fragile and unstable. While one such a crane may be around 30m (98ft) high, those of 80m (262ft), with a lifting capacity of up to 18t (20 US tons), are also possible. The secret of a tower crane's stability is its massive concrete foundation, which may form part of the building's permanent foundations, or be a temporary base removed after the crane's work is done.

Installing this base is the first stage in the deployment of a tower crane. A further mobile crane will be needed to unload all the components from a convoy of tractor-trailer units and begin the assembly of the crane.

Usually two sections of the triangulated lattice mast will be bolted down to the foundation first, topped by a special climbing unit (of which more later). Next comes the slewing unit, an electric motor and gears that rotate the jib.

There are three basic designs of jibs on tower cranes. Often seen on congested city sites are the luffing jibs as used on the Burj Khalifa project. These can swing up and down and have the key advantage of keeping the load and the jib within the confines of the site. Against this is the disadvantage that they are more mechanically complex and slower in operation than the alternatives – the flat-top and hammerhead jibs.

Both of these are horizontal structures built

RIGHT Almost certainly the highest tower cranes in the world once the Burj Khalifa reached its full height. *(Entenman)*

ABOVE Flat-top tower cranes are the simplest. *(Luis Fernandez Garcia)*

up from steel sections. On both, the longer working arms support a trolley from which the lifting cables hang down and can be tracked in and out as desired. The shorter machinery arms contain the main motor (that actually lifts the load), the control electronics and the cable drum and the essential concrete counterweights. There is also a control cabin as well as the very long ladder that is usually the only means of access for the operator.

Flat-top jibs are simple trusses. Hammerhead jibs are more complex structures, on the other hand, featuring a vertical tower above the mast and cable stays running out to both arms to stabilise them when loaded.

All types of tower crane have one really convenient function. They are self-climbing, thanks to the climbing unit installed between the basic mast and the slewing unit. This comes into play when the jib has been assembled, usually on the ground but sometimes in mid-air, and placed on to the slewing unit.

Temporary weights are added to the working arm to balance the permanent counterweights before the jib is lifted off the ground by the mobile crane. When the time comes for the tower to grow, powerful hydraulic jacks on the outside of the climbing unit lift the jib and slewing unit off the tower and raise them 7m (23ft) or so into the air. The crane itself then lifts another 7m (23ft) section of tower into the gap created and the jib and slewing unit is lowered back on to the tower.

As the skyscraper rises, the ground-mounted tower cranes will need extra stability. Cranes located around 2m (6.7ft) outside the building's footprint can be physically tied back to the structure, which must be designed to

take any extra loads generated at this stage. Alternatively, the crane can be founded inside the structure, typically next to a service core, and tied in as before.

Lifting capacity is a function of the distance between the loaded trolley and the central mast. For example, with the trolley 30m (98ft) from the mast, the most powerful tower cranes could lift say 10t (11 US tons). Just 5m (16.4ft) from the mast, however, a 60t (66 US tons) load could be handled. Sensors on the jib and the winch ensure that safe load levels are never exceeded.

Perched so high above the ground, tower crane operators would have difficulty dropping their hooks precisely into the desired location at ground level. Waiting there, however, is an operative dubbed the banksman, who is responsible for the fine alignment of the lift. In earlier days communications between a banksman and crane operator was by means of complex hand signals – these days everything is electronic.

Winds get stronger the higher the crane rises, and as a result become the tower crane's worst enemy. Regulations differ slightly from country to country, but the usual upper limit at which work must stop is a wind speed greater than 20m/sec (65ft/sec). In practice, if the loads being lifted are large and lightweight – such as timber formwork – work usually stops at lower wind speeds as the loads begin to swing dangerously.

Tower cranes are usually hired from specialist companies. Once the skyscraper is effectively complete, the crane must be dismantled and brought back down to ground level – in good condition! This is inevitably a complex operation, but one that very rarely goes wrong.

ABOVE LEFT **Hammerhead cranes are more complex, but more structurally efficient.** *(Axelv)*

ABOVE **Operating a tower crane so far above the ground is not a job for the faint-hearted.** *(Crovax200)*

SLIPPING UP

At the heart of most skyscrapers there will be one or more complex reinforced concrete cores housing the elevators and escalators, emergency stairs and the majority of the services. These cores will be heavily loaded and the steel reinforcement within the concrete will be dense and congested, particularly in the lower sections. At the same time such cores have to be constructed quickly to a very high degree of accuracy.

Concrete sets and hardens through a chemical reaction between the cement and the water in the mix. This reaction is relatively slow. Concrete's compressive strength is usually measured 28 days after placing. However, a purpose-designed mix will develop useful strength in a much shorter time, and advantage can be taken of this to speed construction.

Formwork is the term used to describe the moulds concrete is poured into. Formwork for a skyscraper core is likely to be a complex and expensive metal fabrication, usually encompassing the entire cross-section of the core. Freshly placed concrete, still in its fluid state, can exert considerable outward pressure at the base of the formwork, so for practical reasons the height of any individual pour has to be limited.

Constructing the core well ahead of the floors has several logistical advantages, not least the opportunity to use the innovative jump-lift system. This uses a temporary fixed machine room in the lift shaft that jumps upwards three or four floor levels at a time as the core progresses and the concrete gains enough strength. A bare-bones lift car is installed and used by operatives to access and construct lower floors. When construction is complete, the car is fitted out and finished to normal standards.

Moving workers and materials up and down a skyscraper project is one of the biggest challenges for the contractor. To take the greatest advantage of the jump-lift option, a rapid core construction method is essential. The current alternatives are jump form or slip form construction: both require significant investment in formwork, working platforms and jacking installations. Both also need skilled operatives.

On skyscraper projects both jump form and slip form assemblies will be self-climbing, powered skywards by hydraulic jacks bearing on temporary anchorages in the hardening concrete below the formwork. There will usually be at least three levels of working platforms, with some form of weather protection, not least to the concrete exposed as the formwork climbs. This is vulnerable to drying winds, strong sunlight and sub-zero temperatures, all of which can impair its strength and durability.

Access for the operatives is usually via a trailing stairway hanging from the formwork. There are strict limits on how much weight can be on the platforms at any one time, and provisions must be made for emergency escape.

Jump form and slip form construction differ on only one fundamental issue. In the first, the formwork moves skywards step by step, while the second sees it climb constantly. Constructing a core using jump forms follows a straightforward sequence. Reinforcement steel is fixed, the formwork assembled and the concrete poured. Once the concrete has hardened sufficiently, the formwork is loosened and jacked up to the next level, where the sequence is repeated.

Concreting takes place from the highest platform. One level down is the main working platform, from where steel is fixed and formwork reassembled. The lowest platform hosts operatives who control the jacking and move the jacks up after each pour via temporary anchorages cast into the pour. Once the jacks have moved on, the anchorages are removed and the surface made good.

Horizontal construction joints between pours are usually visible in the concrete. This is rarely a problem as the structural core normally disappears behind internal finishes. Where a higher-quality finish is specified, the alternative is slip forming.

This is a continuous 24/7 operation requiring a highly trained crew, uninterrupted supplies of reinforcement and a high-performance concrete mix that rapidly gains early strength. The formwork is jacked up constantly; the rate of concrete placing is matched to the strength gain of the concrete and the freshly poured concrete is only exposed when it has attained enough strength to be self-supporting.

A climbing rate of up to 350mm/hr (13.8in/hr) is possible. The formwork assembly will be very similar to that used for jump forming, except that steel fixers work on the highest platform. Below them is the concreting platform and on the lowest level skilled operatives operate the jacking system.

With both methods, maintaining the vertical alignment of the core as constructed is another major challenge – traditionally, slip form has been seen as the hardest to keep aligned. These days laser-based surveying equipment has made the task much easier.

OPPOSITE Jump-lift construction can simplify construction logistics. *(Anthea Carter)*

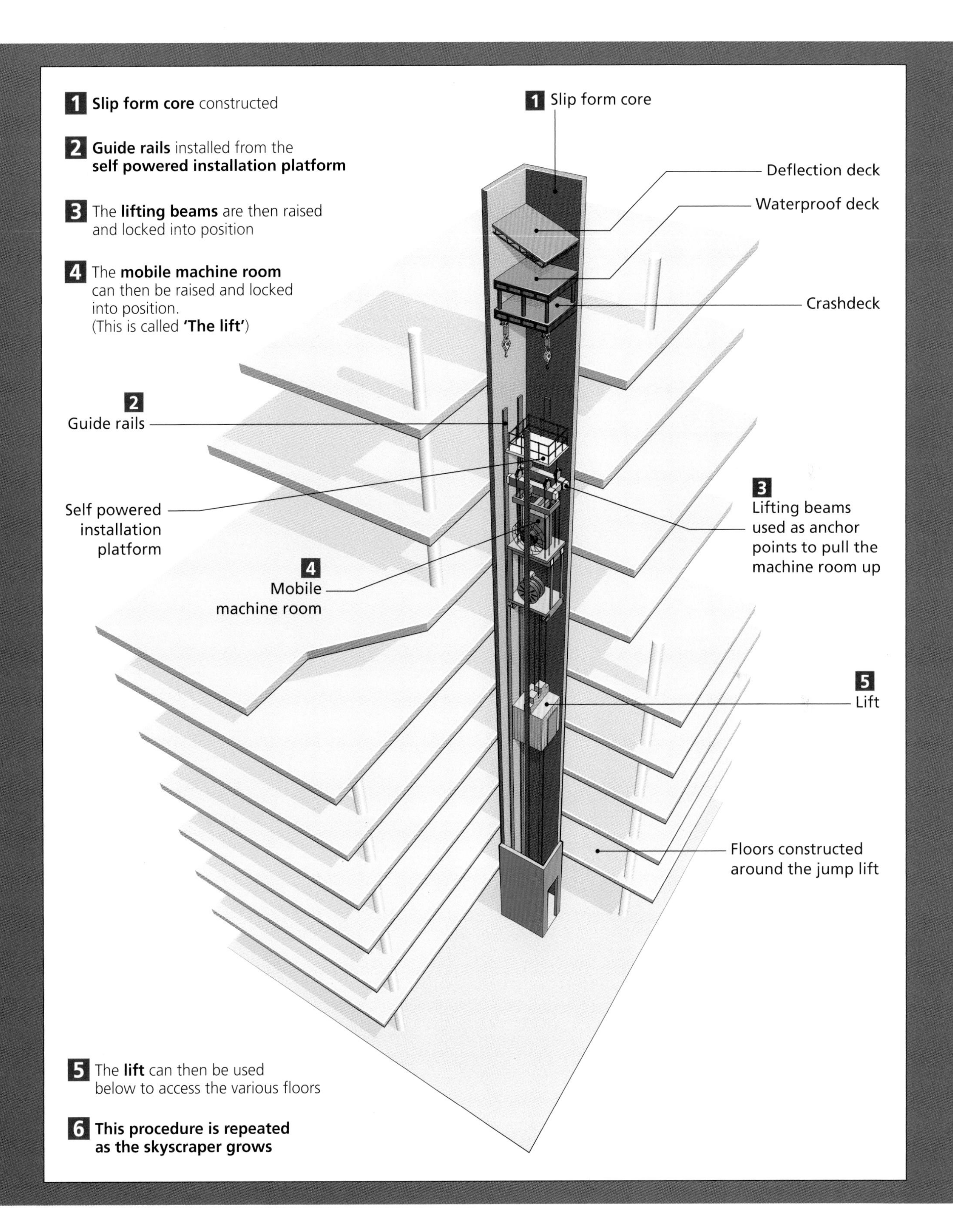
1 Slip form core constructed
2 Guide rails installed from the self powered installation platform
3 The lifting beams are then raised and locked into position
4 The mobile machine room can then be raised and locked into position. (This is called 'The lift')
1 Slip form core
Deflection deck
Waterproof deck
Crashdeck
2 Guide rails
Self powered installation platform
3 Lifting beams used as anchor points to pull the machine room up
4 Mobile machine room
5 Lift
Floors constructed around the jump lift
5 The lift can then be used below to access the various floors
6 This procedure is repeated as the skyscraper grows

They erected the Empire State Building and the original World Trade Center. When the planes smashed into the Twin Towers, they dashed to rescue survivors. They demolished what remained of the structure they had worked on three decades earlier – and were there when the

new Freedom Tower began to rise from the same spot. For more than 80 years there was hardly a skyscraper or bridge project in New York that did not rely on these highly trained teams of so-called skywalkers, famed for their death-defying feats on the high girders. But who were they?

The Mohawk people: these are the skywalkers. Their home is northern New York State and Canada. The story of how they became high-steel skywalkers begins in 1886.

A proposed cantilever railway bridge across the St Lawrence River near Montreal would have one abutment on Mohawk land. In compensation, the bridge-builders agreed to employ tribesmen as labourers – but then found that the young Mohawks took to it.

Soon they became trained riveters, not an occupation for everyone. By the 1920s many four-man gangs of Mohawks were working on Manhattan skyscrapers. These gangs, usually made up of close relatives, followed on behind once the building frame had been temporarily bolted together and aligned. One man would heat the rivets in a portable coal-burning forge until they were red hot and soft, then grasp the rivet in a pair of tongs and toss it to a second man, the sticker-in, who would catch it in an old paint can.

A third man, the bucker-up, had already removed a temporary bolt. The sticker-in then slotted the rivet into the bolthole, the bucker-up braced the rivet with a dolly bar and the riveter hammered it into place with a pneumatic hammer. Roles were regularly switched around, to give the riveter some relief from the bone-shaking hammer. There was soon a sizeable Mohawk community in New York.

Over many decades Mohawk ironworkers were on site of nearly every new skyscraper in the city. Since a building bust in the 1980s their numbers have dwindled.

Today, though, the tradition continues and at least 200 men and women remain involved with the city's construction projects.

One of the Mohawk myths is that they have some inherent physical characteristics that make them uniquely qualified to walk the high steel. The fact that the accidental death rate among Mohawks is much the same as with other high-steel workers would suggest that this is indeed a myth.

Similarly, the Mohawks are not more fearless. They get as scared as anyone else.

LEFT Steel erectors on the Empire State Building took extreme risks. Note the Chrysler Building in the background. *(Public domain)*

PUMP IT UP

In 2009 a new world record was set in Dubai. No, not the topping-out of the world's tallest building, then known as the Burj Dubai (its name changed to Burj Khalifa in 2010). Constructing the supertall skyscraper was only possible with the latest in a relatively new technology – high-pressure concrete pumping.

The high-strength concrete specified for the project was finally pumped over 600m (1,988ft) vertically in a new world record.

As soon as water is added to cement and aggregate, a chemical reaction begins. What starts life as a relatively easy-to-handle fluid mix will soon begin to stiffen and solidify. In practice, the concrete will remain workable for up to two hours on average. High ambient temperatures will significantly shorten this time-frame, while colder temperatures will extend it.

Concrete must remain fluid enough to flow easily around the steel reinforcement, often with the assistance of poker internal vibration, although self-compacting mixes are becoming more popular. If it loses its workability too soon, there will be voids within the hardened concrete and excess trapped air. There are many sophisticated chemical admixtures and additives that can slow down the setting reaction, improve fluidity or increase the final strength – or all three at once. Naturally, these increase the cost of the concrete.

Freshly mixed concrete may be fluid, but it has awkward characteristics. It is heavy (2.5 times denser than water), it contains hard rock particles that make it highly abrasive and there is always the risk that unplanned delays could lead to the concrete beginning to set hard while still in the pump.

High-performance concrete pumps use a twin-piston layout, which can produce the high pressures needed to force the concrete to great heights. On the world record pour at the Burj Dubai/Khalifa, pump pressure peaked at 200 atmospheres. It took the concrete around 40 minutes to travel from the pump delivery hopper to the pour more than 600m (1,969 ft) above, at which point there was around 26t (28.6 US tons) of concrete in the delivery pipeline. Some 28m^3 (989ft^3) of concrete per hour was delivered by very large pumps powered by massive diesel engines.

RIGHT Concrete needs to remain fluid while being pumped on site. *(Shutterstock)*

People movers

For all their monumental glory, skyscrapers are created as spaces to be occupied by people. All the detail that goes into considering structural form needs to be matched by considerable forward planning for how occupants will move between so many floors.

Put simply, stairs become impractical as the sole provider of vertical access for buildings with more than a few storeys. Escalators – moving stairways – are the obvious next evolution. However, in skyscrapers they typically only serve the lower floors, as evidenced by the familiar sight of escalators in multi-use buildings with retail space and mezzanine lobby spaces.

The development of the commercially viable elevator in the last half of the 19th century unleashed skyscraper potential and, ever since, each one has required banks of elevators for the occupants' convenience. These normally take the form of clusters of vertical shafts, each housing a single counterbalanced cabin suspended from steel cables and powered by an electric motor.

In practice, the type and number of elevators and escalators can vary greatly from one high-rise to the next. Choosing the right combination of transit options is governed by several key considerations, not least the number of users to be served. Whether moving stairways are the right answer for a given

ABOVE Escalators are a practical way to access lower floors of skyscrapers. *(Shutterstock)*

BELOW Choosing the right combination of transit options is determined by the predicted number of occupants moving up and down a skyscraper at different times of day. *(Shutterstock)*

ABOVE **Office workers account for the highest number of occupants of the world's skyscrapers.** *(Shutterstock)*

space is determined by the number of floors needing access as well as the function of the internal space. It could be office, residential, hotel, retail or fitness facilities, each providing the building designer with a valuable indication of how many people will need access to a floor at any given time of day.

The most popular occupants of the world's skyscrapers are office workers. In these buildings peak-time traffic usually hits around the middle of the day as workers down tools and head out for their daily lunch break. Unlike the one-way rush-hour traffic at the beginning and end of the day as people arrive for work and depart for home, this midday traffic is more complex. Workers are moving in and out and up and down many floors, all in a relatively short period of time.

By contrast, hotel and residential occupants usually contribute to peak usage throughout the evening, again with a mix of two-way traffic.

Meanwhile, the challenge for skyscraper retail space is to ensure shoppers are afforded access to the maximum number of outlets with the greatest of ease, whether that is a multi-level mall, a single department store or a cluster of convenience and food retailers.

OPPOSITE **Buildings above 50 storeys result in much longer travel times with single-cabin elevator shafts.** *(Shutterstock)*

Density of occupation is also a vital consideration that needs to be successfully anticipated and planned for. The number of people occupying a certain area varies across the globe, along with a move away from enclosed internal offices towards the inclusion of more open plan 'hot desk', high-density spaces can make that job more challenging.

Similarly, luxury residential units will be more capacious and generous in their allowance of space per resident, while more affordable housing family units will assume more people inhabiting the same space. The same variations apply to the range of affordable to high-end hotel accommodation available.

Whatever the variables, installing the right number of elevators is critical. The cost implications can be huge when prime real estate on offer is diminished greatly if too large a floor area is taken up by overestimating numbers of elevator shafts. Equally, unhappy inhabitants having to wait a long time for too few elevators will also take its toll on leasers of premium space.

The range of elevators has begun to expand in recent years following decades where most operated with very small advancements on the mechanisms that served the earliest skyscrapers. Since the start of the 21st century, innovations have sought to boost efficacy and speed to keep pace with the increasing heights and changing geometry of skyscraper design.

Beyond 50 storeys, single-cabin shafts become less and less effective. Passenger travel time increases significantly. There are only so many gains that can be made by superfast elevators capable of around 18m/sec (60 ft/sec) and increasing the number of single cabins in single shafts would make for highly inefficient floor plans on lower floors, which need fewer elevators for optimum circulation.

Sky lobbies have contributed a great deal to solving this conundrum. These are intermediate transfer floors where high-speed elevators stop only every 50–100m (164–328ft). At these landing points passengers can change from an express to a local elevator to access the floors in between.

Stacking the local cabins in their own shaft, one above the other from the lowest level to the upper levels, is the most efficient way to do this, leaving express shafts free for higher-speed cabins. Generally speaking, this minimises the floor area taken up by shafts,

continued on page 106

TESTING TOWERS

With the safety of elevators being so critical, you may wonder how the manufacturers test new and improved elevators before bringing them to market. There really is only one answer – it needs a tall building for manufacturers to practise in. An actual building under construction is hardly an ideal venue for a meaningful test programme. So all that is left is to build a bespoke elevator-testing tower.

For the largest manufacturers, this has become an architectural art form in itself – and an excuse to outdo the competition.

Usually around 100m–200m (328ft–656ft)-tall, the main purpose of elevator-testing towers is to examine the stress limits and wear and tear of cabins and their mechanisms within a controlled environment. This is done to prove the capacity and safety of current designs and, importantly, to resolve any issues before going into full-scale production.

This may be the obvious primary function, but the 246m (807ft)-tall, 21m (69ft)-diameter Rottweil Test Tower in Germany – constructed by international elevator supplier ThyssenKrupp – is becoming something of a star attraction in its own right.

The twisting concrete tower is palely clad in an unusual mesh that gleams as it reflects the light throughout the day.

Not only does it conduct the latest high-tech elevator tests; since it opened in 2017 it has become a draw for sightseers and boasts Germany's highest visitor platform at 232m (671ft). It even has conference and meeting facilities for hire and plays host to a tower-running event for super-fit competitive athletes.

One of the tallest and most striking of its type, the tower features 12 shafts that can test cabins travelling at speeds of up to 18m/sec (59ft/sec).

OPPOSITE The 246m (807ft)-tall Rottweil elevator-testing tower in Germany not only provides a research and development base for new designs but is also a visitor attraction and conference facility. *(Shutterstock)*

RIGHT The twisting concrete tower is clad in a gleaming mesh that reflects the light. *(Shutterstock)*

but even so, on some buildings this can still be as much as 40% of the floor plan.

Manufacturers have continued to push the limits and double-deck cabins have been introduced in recent years. These allow passengers for two floors to be served simultaneously, creating extra capacity without the need to add more space-grabbing shafts.

There are limitations to both. Sky lobbies introduce a not-so-convenient interchange for local-floor passengers, something that becomes an irritation for those wishing to reach the uppermost intervening floors. And although these cope well with the peak uses of, say, an office building, where most traffic goes in one direction, serving the complex two-way lunch rush causes a dramatic decline in any time-efficiency gains.

The emergence of variable floor heights in tall buildings has blocked double-deck elevators from becoming a runaway success. Clever mechanical interventions mean the two connected and stacked cabins can adjust to edge closer together or shift further apart to allow for this quirk. But this adds extra stoppage time and makes their widespread use less appealing to building owners seeking the speediest passenger transit times.

Hoping to solve this is a recent manufacturing development that is a variation on the double-deck design, albeit one that is technologically more complex. This innovation enables two independently controlled elevators to occupy a single shaft, generating all the space-saving of double-deck elevators, while being flexible enough to transport more passengers precisely where they need to be in the fastest time possible.

In this system, both cabins use the same rails within a shaft but rely on separate motors and counterweights. They are supported by a unique roping arrangement that enables them to move independently.

The 185m (606ft)-tall European Central Bank building in Frankfurt was completed in 2014

LEFT Recent developments have seen the introduction of double-deck and multi-cabin shafts to improve passenger travel times within the tallest skyscrapers. *(Chris 73)*

and has seven of these independently moving, double-deck elevators for its two towers, supplemented only by single conventional and separate goods elevators that would also evacuate occupants in the event of a fire. It also has five express elevators in its atrium.

To ensure effortless and efficient circulation, terminals at each landing allow passengers to programme their floor destination. This system is now widely in use and is increasingly favoured over the conventional two-button control panel. Software is able to compute the quickest connection between the floors and digital displays direct the passenger to the chosen elevator.

Even greater innovations are on the horizon that could transform skyscraper transit with the development of rope-free technology. If successfully proven and commercially viable, traditional cables and counterweights with electric motors on top of the shafts will be an outdated concept.

Cabins would instead be equipped with their own induction motor and could be fabricated from high-strength but lightweight composite materials that could generate significant space and cost savings. Without the restrictions of cables, multiple cars could operate within in a single shaft, moving independently in response to passenger demand.

Current direction restrictions on movements will be lifted, and horizontal or diagonal transits will become just as achievable as today's vertical elevators. It is hoped this would mean cabins could reach passengers at landing levels as often as every 15–30 seconds.

Use of this type of system in buildings above 300m (984ft) could reduce elevator footprints by as much as half, and research and development continues in a specially built testing tower in the hope that the technology will become viable at a broad commercial scale.

A similarly enticing development is the idea of an articulated funiculator system. While different from the rope-free option with its continued reliance on cables, the funiculator would represent another leap forward for elevator design.

Consisting of a network of elevating trains with up to four pod-like carriages that circulate through two connected shafts, they would stop only at sky lobby transfer stations, usually at least 100m (328ft) apart. These stations may be vertical, as with traditional sky lobbies, or they may be positioned on a looping shaft in a horizontal direction, as both horizontal and diagonal motions would be possible.

Freeing skyscraper designers from factoring in vast areas for shafts would allow them to generate the same or more usable floor space over less land take. Yet again, elevator technology may yet be responsible for great advancements in tall building design and could see a surge in the trend toward slimmer profiles, a tantalising prospect that benefits developers with the associated cost savings on façade and land costs.

One of the world's scariest jobs

Skyscrapers are designed to withstand the effects of nature and general daily use by their occupants with as little as possible maintenance intervention. Key elements are hidden or protected from the elements: for example, a steel structural frame will be sheathed in fire protection materials and located inside the building.

Similarly, foundations are buried deep into the ground. And while modern engineering techniques have emerged to help deal with issues if they arise, such as adding new piles via a technique called underpinning to deal with subsidence, there is usually little need for maintenance. Foundations are designed to be robust enough to support the building throughout its working life.

However, the vast façades of a skyscraper bear the full brunt of nature's assaults.

As a consequence, much effort has gone into ensuring that the curtain wall performs well. This non-structural cladding forms the external walls of a skyscraper – often combining a metal frame with glass panes – and performs a number of functions, not least keeping the weather out and the inhabitants in (*see* Life in the sky, p.49).

The façade still needs constant attention, most routinely in the form of window cleaning. This apparently menial task is by comparison to the domestic chore a rather more terrifying

ABOVE BMUs have platforms that are mechanically operated from the roof of a tall building. *(Shutterstock)*

prospect and is a job suited only to a very small proportion of the population.

Early skyscrapers had windows that opened from the inside and often had ledges that maintenance and cleaning staff could climb on to before securing themselves with harnesses on to the window frames. While they still needed a head for heights, logistically the work was relatively straightforward.

All that changed with the development of modern curtain wall systems. Importantly, these completely seal the inner building.

BELOW Even very simple tasks become more complex at great heights. *(Shutterstock)*

Despite their use being well established over the past century, consideration for how easy it will be to clean and maintain the façade is still often given too little attention by skyscraper designers.

The most popular equipment needed to clean and take care of the external walls of a modern high-rise is its building maintenance unit (BMU). This has two key parts. The first is a load-bearing structure mounted on tracks on the roof, perhaps with additional units installed on external lower levels where a building has a stepped-back design. The second is an enclosed platform that is mechanically operated and can be lowered and positioned around the building by bearing ropes on the first part.

From this platform, highly skilled and tethered professionals can carry out the exhaustive window cleaning as well as some maintenance, such as the replacement of heavy glass façade elements.

Even the seemingly easy part of window cleaning takes on a new level of complexity at skyscraper scale. While the detergent, rubber squeegee and cloths may look familiar, gravity and wind play a major role and dropping a single small cleaning utensil from an extreme height can have serious implications.

A careful approach to tethering the necessary equipment correctly and a diligence throughout the process is vital.

The effects of wind are also exacerbated at greater heights and have to be given considerable forethought. Wind speed is monitored by operatives to ensure they do not take undue risk. This is particularly true where access via a BMU platform is not possible. Many skyscraper designs have in recent years eschewed linear angles in favour of more interesting and complex geometric designs. Classic BMUs are no longer feasible in those cases and so abseiling climbers are the only alternative.

Of course the margin for error at height is almost non-existent, so the accredited training that is undertaken by these specialist professionals is meticulous. Safety is paramount and drills and checks are performed regularly to ensure all are up to date to prevent devastating accidents.

Robotics and technology may one day all but eradicate the dangers of this work. Window cleaning robots that are able to scale and clean façades and windows are increasingly commonplace. Elsewhere, research into the capacity of flying robotics, or drones, to take on the majority of these tasks is gaining momentum.

Safety is a great driver for these changes to be brought about. However, the more likely push to move towards machine-only maintenance will take hold when the technology is proven to be the most capable, efficient and cheapest means of dealing with the skyscraper maintenance challenge.

The technology is there for robotics to undertake relatively lightweight cleaning tasks, but experts agree there is still a way to go before there is proven capacity for safely completing more heavy-duty maintenance, such as façade panel or glass pane replacement.

While there remains no viable technological solution for carrying out this work from the

RIGHT Abseiling climbers make it possible to clean buildings such as the Burj Khalifa when unusual geometric building designs make it impossible to rely on mechanical platforms. *(Shutterstock)*

ABOVE Robotics and technology may one day eradicate most of the dangerous work from skyscraper maintenance. *(Shutterstock)*

outside, building designers need to instead give even greater care and consideration to a building's maintenance needs up front. It is possible to design façades that make it possible to replace the glazing from inside, but this requires clever thinking about how this will work from the very earliest stages.

Dealing with disasters

Skyscrapers are resilient buildings. Disasters are thankfully rare, even though there are a number of threats to their existence that must be considered during the design phase. In terms of the likelihood of extreme events occurring, the most likely is fire, followed by aircraft impact then terrorist bombs. Earthquakes are less of a threat than may be imagined, while responsible designers must now take seriously the potential risks associated with climate change.

Resisting the flames

Skyscraper fires may be rare and catastrophes even rarer, but every skyscraper is designed from the outset to resist any major blaze for at least two hours. This is to allow time for the occupants to evacuate safely and for firefighters to enter the building and begin dowsing the flames.

A building designer must first consider the main structural frame. This will be either steel, concrete or a hybrid of both – although the day of the engineered timber skyscraper may soon be dawning. Each material reacts to fire in its own unique fashion and needs to be considered on its individual merits.

In many ways steel is the ideal structural option – but it does have an Achilles heel. Unprotected steel will start to soften at around 450°C (842°F) and will have lost half its strength by the time the temperature tops 650°C (1,202°F). Structural steel under load will sag, twist and buckle. Fires in both office and residential areas will reach temperatures well in excess of these critical levels in remarkably short time periods. This being the case, the steel needs protection if premature collapse is to be avoided. Achieving the 120-minute level of protection normally specified for buildings more than 30m (98ft) tall can be challenging.

Encasing the steel in a reinforced concrete shell was one of the earliest techniques, and one that is still often used in basement car parks and the like where accidental impacts on the columns are likely. This is not a realistic option for the rest of the building, however, due to the added weight involved. However, there is a 'concrete' option, more popular in the USA than elsewhere, that is particularly quick and easy to apply.

Originally this was a fluid mix of Portland cement, asbestos fibres, sand and water, spray-applied to the structural steelwork after erection but before internal finishes were installed. This technique could cope with complex steel sections and joints and was relatively cheap. Modern formulations are asbestos-free.

However, skill is needed to achieve a consistent coating thickness, and there is always the risk of overspray. Focus shifted away from this type of coating some decades ago as the new technology of intumescent coatings developed.

These are paint-like materials with a complex chemical composition that reacts to rising temperatures as a fire develops. At around 200°C (392°F) a dramatic swelling will begin as a chemical reaction is triggered. The result is a layer of insulating char up to 50 times thicker than the original coating, for instance a 1mm (0.04in)-thick intumescent coating will expand to a 50mm (1.97in) thickness to protect the underlying steel.

OPPOSITE Fortunately, skyscraper fires are rare and casualties even rarer. *(Shutterstock)*

Intumescents can be applied on site, although steel sections pre-coated in the factory are also available. Until recently 'thin film' intumescents struggled to achieve 120 minutes' resistance, but now there are products for which such performance is guaranteed. For extreme exposure conditions or areas where maintenance would be very difficult, there is a range of 'thick film' intumescents.

These were developed to resist hydrocarbon fires in offshore oil installations and are very durable. They share with their thin-film cousins the potential to have decorative topcoats applied over the reactive base coats.

Steel sections can also be protected by fire-resistant boards. Fixing these is a dry trade, which may be preferred for logistical reasons as it is usually less disruptive than wet trades such as spray-applied coatings. The advantages also include consistent thickness (hence uniform fire resistance), and in the case of heavyweight boards, offer a clean, boxed appearance, receptive to decorative finishes.

Cheaper lightweight boards are usually preferred where aesthetics are unimportant. In such areas a recent option is flexible blanket systems, which are easier to apply than boards on complex structures. Again, this is a dry trade with minimal impact on logistics.

Structural reinforced concrete is usually regarded as 'fireproof', and in most applications this is effectively true – a concrete structural element will normally achieve the necessary 120-minute fire resistance without difficulty. There are two problem areas, however. Concrete is a very variable material. There are literally thousands of combinations of cement, aggregates, chemical admixtures and water, each of which will react a little differently in a fire situation. Only a large-scale laboratory fire test can give a reliable indication of how a particular mix proposed for a particular project will perform.

For normal-strength concretes, such tests are usually superfluous, unless, perhaps, if an unusual cement or aggregate is proposed, possibly for aesthetic reasons. High-strength concretes are a different matter. As concrete technology has developed, much higher-strength concretes have become almost routine on many high-rise projects. Beams and

columns can be slimmer and lighter, reducing the loads on foundations and freeing up more internal space. There is a very good economic argument for going down this route.

High-strength concretes are much less porous than normal mixes but still contain a small percentage of unreacted free water. As the temperature rises in a fire, this water turns to steam and tries to escape. A normal-strength mix has enough pores for this to happen safely; in the dense, low-porosity, high-strength mixes, however, the steam is trapped, pressure builds up and eventually areas of the concrete face will be forced off.

This 'spalling' can be violent and explosive. The loss of significant amounts of concrete might compromise the load-bearing capacity of the structural element. However, there are ways of mitigating this tendency, such as the inclusion of plastic microfibres. Those near the surface soon melt, leaving behind cavities through which the steam can escape.

Once the building designer has finalised the fire protection of the structural frame, attention must turn to how occupants will be protected in a fire situation. Building codes and regulations across the world will set similar standards for maximum compartment size, protected escape routes, firefighter access and the like – but there is no universal standard covering how occupants should evacuate a building should it become essential. In the case of a skyscraper, this is one of the trickiest challenges to face the designer.

Emergency escape stairs must be fire protected, and on most sub-100m (328ft)-tall buildings they will share a single central concrete core with the elevators. There are regulations covering the dimensions of the stairs and of the fire doors giving access to them. Emergency stairs have to have smoke-control measures as well, such as forced ventilation or pressurisation – the latter keeps the pressure inside the stairwell higher than the rest of the building.

However, should everyone in the building head for the escape stairs immediately after hearing a fire alarm, chaos can ensue.

People have been trampled to death in such situations. To avoid similar tragedies, modern skyscrapers usually have procedures for phased evacuation, where the floors closest to the seat of the fire are evacuated first, followed in orderly fashion by the remainder in order of perceived risk.

Unfortunately, unless such an evacuation is well rehearsed and familiar to occupants, there is a strong possibility that there will be a universal stampede towards the escape stairs, with disaster almost inevitable. There is no way a functional skyscraper can be designed with escape stairs large enough to cope safely with a panic-stricken simultaneous evacuation from all floors.

One option is the so-called 'defend in place' strategy. Most often chosen for residential/hotel accommodation, this is based on the compartmentation of the building. Each individual apartment or hotel room will be enclosed within fire resistant walls, floors, doors and ceilings. Should a fire break out, occupants not directly threatened immediately are advised to stay put and wait for firefighters to extinguish the blaze and give the all-clear.

Skyscrapers with open-plan office floors can have fire-resistant refuges on individual floors where workers can wait for the all-clear. The great advantage of 'defend in place' is that it leaves the emergency stairs clear for the firefighters. In other scenarios the firefighters will have to force their way upwards against a flood of panicking escapees.

Recent supertall skyscrapers are increasingly likely to have included several 'fire break floors' incorporated into their design. These are normally dedicated mechanical plant floors with a low fire load that effectively divide the building into discrete sections. This allows the fire protection and evacuation options in each section to be designed separately, a particular benefit for multi-functional buildings that may combine residential, office and hotel functions.

Tall buildings will usually have a dedicated firefighters access stair, which the emergency services can enter directly from outside the building, and even a dedicated elevator, again with effective smoke-control measures. These will usually share the core with the public emergency stairs.

Despite formal warnings against the use of elevators in a fire situation (if the fire cuts off the power, anyone in an elevator may be trapped), lessons learned from the 9/11 tragedy (*see* Lessons from 9/11, p.177) suggest that elevators can be a vital escape route during a mass evacuation. This is particularly true when disabled or elderly occupants are involved. For such less fit and mobile individuals the challenge of descending more than 100 flights of stairs would be beyond their capacity, and an elevator would be the only way out.

Since 9/11 a number of alternative evacuation methods have been put forward. These include parachutes, escape pods on zip wires, external emergency elevators and external ladders. None are currently seen as realistic options.

Helicopter rescue may seem like an obvious alternative, but there are serious problems associated with it. Although many tall buildings were originally constructed with helipads on their roofs, their main function was to allow chief executives to avoid traffic jams. In a fire situation, the smoke and turbulence created by updraughts of hot air would seriously hamper helicopter evacuation efforts. In any case, the passenger capacity of typical emergency services helicopters is limited, as is their number, so even in ideal conditions they could make only a minor contribution to a mass evacuation.

Building codes and regulations have to be both prescriptive and conservative. Their purpose is to save lives, not property. In themselves they do not guarantee that a particular building will perform as predicted by its designers. Until recently, however, designers have had few alternatives but to stick to the codes, and on the whole this approach has worked.

With the increasing availability of powerful computers, a realistic alternative is now accessible. Performance-based fire safety design, better known as fire engineering, allows designers to model how a virtual blaze will develop and how the virtual building and its occupants will react, thanks to techniques such as CFD and software that can predict the behaviour of crowds in stressful situations.

Initial design proposals can be tweaked until an effective solution is achieved. This may well result in less fire protection being specified for areas of the building unlikely to be seriously affected in a fire situation. Modelling how changes to escape routes, automatic sprinkler systems and public address systems can significantly reduce evacuation times is even more important.

Raining indoors

Sprinklers work. Research has shown that properly fitted and maintained automatic sprinkler systems have reliability and effectiveness ratings close to 100%. It is also claimed that in the UK no one has ever died as a result of a fire in a building fitted with sprinklers.

So effective are they that regulations allow other fire prevention and resistance measures to be relaxed. Compartments can be larger, tall buildings can be built closer together, design can be more flexible. In modern skyscrapers, be they office blocks, hotels or residential (or a combination of all three), sprinkler systems are almost universal.

Yet they were slow to be accepted, particularly by office developers. Much of this reluctance appears to be the widespread belief, fostered by too many irresponsible movie and TV scenes, that a small fire in one location will automatically trigger every sprinkler on every floor of the building. Occupants will be soaked, essential paperwork saturated and expensive electronic equipment ruined.

The truth is that only an individual sprinkler in close proximity to a small fire will react to the heat from that fire and start to spray water. Sprinkler heads further away from the blaze will be unaffected.

A basic sprinkler system consists of a header tank containing mains water, a

BELOW There is widespread misunderstanding of how automatic sprinklers actually work. *(Brandon Leon)*

distribution system and an array of sprinkler heads mounted in the ceilings. Typically, in the heads will be liquid-filled glass bulbs holding a plug in place. When a bulb is subjected to localised heating, the liquid will expand, fracture the bulb and release the water.

Some people dislike the industrial appearance of a standard sprinkler head. However, decorative alternatives are available and are perhaps more suitable for residential areas.

Regulations vary from country to country, but an automatic sprinkler system in a skyscraper would typically have to have the capacity to supply enough water for the sprinklers to operate for 90 minutes. This should be enough time for firefighters to arrive and deal with any persistent flames.

A more recent alternative is the water mist system. These use high-pressure pumps to force water through special nozzles that produce very fine droplets. A mist is formed, which cools the flames and deprives the fire of oxygen. Originally designed for high-risk industrial applications, water mist systems have the advantage of requiring much less water than sprinklers to achieve the same objective.

Less water should mean restricted water damage to property and reduced the mess to be cleaned up. The fine droplets can even 'wash' smoke from the air, minimising the spread of smoke within the building.

Such systems are most effective in compartmentalised buildings and are less so on open-plan office floors and the like. Air conditioning or draughts from open doors and windows can actually blow the mist away.

Terror from the sky

Aircraft rarely crash into skyscrapers. Most of the recorded incidents involve small private planes, and fatalities are thankfully low. Almost all impacts are accidental, usually caused by poor visibility, although there are a handful of cases where the collision appears to have been deliberate. Suicide seems to be the main motive, although there is one unique event where fanatical terrorists aimed to create as much death and destruction as they could manage.

This was the 9/11 World Trade Center disaster in 2001. Two hijacked Boeing 767 airliners, each weighing around 150t (165 US tons) and with some 38,000 litres (10,000 US gallons) of jet fuel in its tanks, were deliberately crashed into the upper levels of the Twin Towers at around 750kph (470mph). Massive structural damage resulted – but the towers stayed upright. It was the subsequent fires that eventually caused their shocking collapses (*see* Lessons from 9/11, p.177).

Only one other incident remotely resembles 9/11. In July 1945 a twin-engined B-25 Mitchell bomber persisted in a landing attempt at Newark Airport in New Jersey despite being warned of zero visibility over the area due to dense fog. Disoriented by the conditions, the pilot turned right instead of left and crashed

RIGHT This B-25A is only a fraction of the weight of the Boeing 767s that struck the World Trade Center. *(USAF)*

into the north façade of the Empire State Building between the 78th and 80th floors.

Despite weighing only 10t (11 US tons) and travelling at little more than 241kph (150mph), the bomber punched a hole measuring around 5.5m by 6m (18ft by 19.7ft) through the façade. One of the B-25's 1,800hp Wright Twin Cyclone engines crashed right through the building and emerged from the south façade. The 1,000kg engine still had enough impetus to reach the roof of a nearby (much lower) building, where it started a fire.

Fire broke out around the point of impact as the bomber's fuel tanks exploded. This was extinguished in less than 40 minutes. Eventually it was discovered that three crewmen had died in the crash, along with 11 occupants of the tower. One female elevator operator had a miraculous escape when the second engine and part of the landing gear smashed into her elevator shaft, shearing the cables.

Otis safety brakes on the elevator car should have locked it safely into place. Unfortunately, the brakes were not designed to cope with more than a ton of aircraft debris landing on the roof of the car. The elevator plummeted 75 storeys to the basement, but Betty Lou Oliver survived, despite sustaining injuries.

It appears that the brakes did slow the car significantly, while the severed cable fell much faster and piled up in the basement, where it acted as a life-saving shock absorber.

Amazingly, despite the tragedy and the destruction, most floors in the building were open for business only two days later. It was 13 years before One World Trade Center opened on the site of the original Twin Towers.

Terrorist attack

Over the last few decades terrorists of various persuasions have turned to the truck bomb as an effective weapon, capable of causing enormous economic damage and putting pressure on governments. A truck can carry much more explosive material than a rucksack or even a car. A large bomb made up from relatively low-grade explosives can do as much damage as a small device containing military-grade high-explosive. The latter is expensive and hard to obtain, while so-called 'fertiliser bombs' can be concocted by almost anyone anywhere.

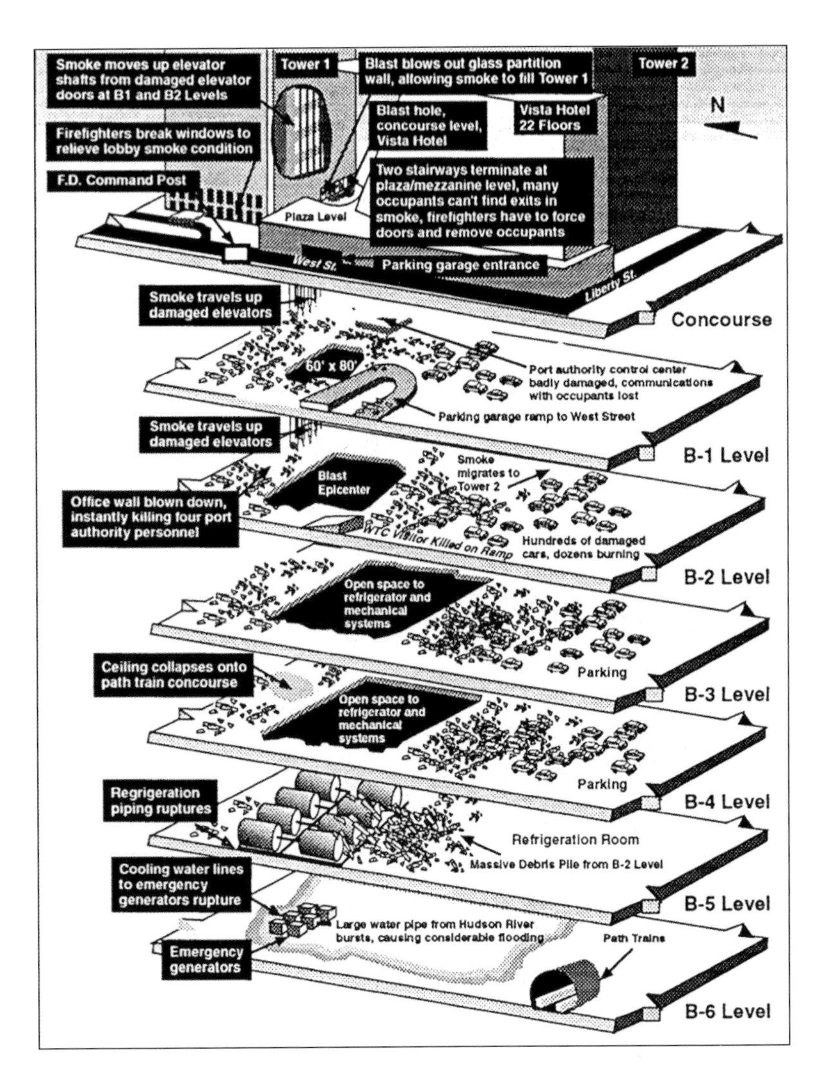

A favourite weapon of the Irish Republican Army (IRA) who used it to attack London on several occasions, a home-made truck bomb was also used by the first terrorists to attack the World Trade Center, back in 1993. Islamic terrorists parked a truck containing a complex home-made 600kg (1,322lb) explosive device in the public basement car park below the North Tower, lit the fuse and fled.

They hoped the bomb would topple the North Tower into the South Tower, bringing them both down and killing up to 250,000 people. In the event, the tower absorbed the shock of the explosion with relative ease. Holes were blown through the basement floors and the main electrical supply was knocked out, along with the emergency lighting. Occupants were, however, trapped in elevators, and thick smoke hampered rescue operations.

Six people died and more than 1,000 were injured, mostly from smoke inhalation.

ABOVE Terrorists hoped to topple the World Trade Center with a truck bomb. They failed. *(US Fire Department)*

Evacuation was slow. Scores of people took refuge on the roof, but only 28 with obvious medical problems could be rescued by helicopter.

Skyscraper design, indeed the design of all buildings that might become terrorist targets, now features a number of measures that will prevent any form of vehicle bomb getting close enough to cause significant damage. Stringent security checks at the entrances to major buildings seem to have reduced the risk of backpack bombs, although the risk from the suicide bomber is harder to eliminate.

BELOW This 14-storey building in Anchorage, Alaska, suffered significant damage in the 1964 earthquake. *(Public domain)*

When the earth moves

Skyscrapers can be found in earthquake-prone locations all over the world. Los Angeles and San Francisco spring to mind, along with Hong Kong, Tokyo and Shanghai. Indeed, any skyscraper located close to a known fault line runs the risk of being hit by a major tremor at some point.

Looking up at a slender 100-storey-plus skyscraper, it would be easy to imagine that it would be terribly vulnerable should the very earth below it begin to shake and heave. Yet, paradoxically, the safest place during an earthquake can be inside a skyscraper. It is medium-rise buildings, up to 20 storeys or so, which can suffer the worst damage during a major quake.

This is down to the way the ground moves during a typical seismic event. It will shake sideways and vertically, with the sideways movement usually much greater. Most buildings will absorb the vertical shaking without much trouble: after all, they were designed to take vertical loads. Horizontal shaking is much more difficult to deal with.

Every building has a natural frequency of oscillation. As the earth moves backwards and forwards at a particular frequency, the building will respond. Should the earthquake frequency be the same as the building's natural frequency, resonance will build up and amplify the sway. If the tremor lasts long enough, the sway can become excessive, and serious structural damage or even collapse can occur.

Luckily, the natural oscillation frequency of skyscrapers is normally much slower than the typical frequency of the horizontal seismic movements, so damaging resonance is unlikely to occur. Measures to control wind-induced sway will usually be more than adequate to deal with earthquake sway.

As skyscrapers rise to unprecedented heights, however, it becomes ever more difficult to predict how a supertall or megatall tower might react in a major quake. A number of ingenious structural systems have been developed to cope with this uncertainty – and more are sure to follow.

There is one side effect of the seismic shaking that can be dangerous, although it mainly affects low-rise structures. Certain soils,

ABOVE Rising sea levels and more extreme weather events will threaten basements in years to come. *(Jerry Stratton)*

usually those containing a high percentage of water, can liquefy when shaken at the right frequency. Any buildings founded on such soils will actually sink into the ground during a quake.

These soils are easily detected during a preliminary site survey and would normally not stop a skyscraper from being built there – its foundations would simply be taken down to bedrock below the suspect strata.

Climate change

No reputable tall building designer can ignore the potential risks from climate change. Many cities are built alongside major rivers or estuaries, or right next to the sea. Skyscrapers in such cities are likely to be more at risk from floods, storm surges and tsunamis than from earthquakes, terrorism or typhoons. As extreme weather events become more frequent and sea levels rise, these risks can only increase.

Skyscrapers typically have several basement levels, housing all their vital plant and equipment. Should water flood into these basements and knock out the electrical supply, the building would be crippled. Elevators will stop working, as will the heating and ventilation systems. Lights will go out, internet connections will fail. For an indefinite period the skyscraper will be virtually uninhabitable and non-functional. The costs involved in repairing/replacing all the sodden equipment in the basement could be staggering – as could the flood risk insurance premiums for skyscrapers in vulnerable areas.

Fortunately, there are a number of proven flood-protection technologies available. These include automatic flood barriers at all ground-level entrances into the basement in particular and the ground floor generally. Where underground services – electricity, landlines, internet – enter through the basement walls, more barriers are essential. There must also be precautions against the sewerage system backing up and heavily contaminated water flooding into the basement.

Chapter Four

Landmark skyscrapers

As skyscrapers evolved from the late 19th century onwards, there were some that became iconic landmarks that had a profound influence on tall building design for decades to follow. Here we look at nine such buildings, ranging from the first ten-storey 'skyscraper' to the current world height record holder.

OPPOSITE Viewed from the Empire State Building, the Chrysler Building is obviously the loser in the 'Race to the Sky'. *(Misterweiss)*

P
A
R
K
STADIUM
AND
PUBLIC
PARKING

Wainwright Building, 1891

The terracotta tower – and the celebrated architect who designed it

Opened in 1891 and still standing in downtown St Louis, Missouri, the ten-storey Wainwright Building has been officially described as 'a highly influential prototype of the modern office building' and 'the first skyscraper that truly looked the part'.

Its architect, Chicago-based Louis Henry Sullivan, was later dubbed the 'Father of Skyscrapers', although in practice he designed very few. Nevertheless, his design principles influenced many other US architects, creating what became known as the Chicago School of tall building design in the late 19th century.

Sullivan's guiding principles are fully displayed on the Wainwright Building. As was the norm at the time, the 45m (147ft) tower's floors rise directly from the sidewalk. To emphasise the height of what was then seen as a very tall building, Sullivan created a tripartite, multi-function design modelled on a classic Roman or Ancient Greek column.

Shops directly accessible from the sidewalk occupied the ground-floor 'base'; the 'shaft' was office space; while the 'capital' was the top-floor attic space housing water tanks and elevator machinery.

Structurally, however, the building broke with traditional and classic construction techniques. An innovative load-bearing, all-steel structural frame allowed for significantly larger windows than would have been possible in load-bearing walls. This was particularly obvious at ground level, where the retail spaces could have the grand windows taken for granted today.

Sullivan also turned away from the plethora of ornate stone carvings and statues that adorned many neoclassical buildings. Instead, he clad his building with lightweight terracotta panels, many embellished with complex mouldings.

Well received by critics and the public alike, the Wainwright Building was also a commercial success. Designation as a National Historic Landmark saved it from demolition in 1968. Missouri state offices now occupy it.

Of mixed Irish and Swiss descent, Sullivan was born in Boston, Massachusetts, in 1856. A precocious student, he left high school a year early and entered the Massachusetts Institute of Technology aged just 16.

OPPOSITE Still standing in St Louis, the 1891 Wainwright Building is seen by many as the prototype of the modern office block. *(Public domain)*

LEFT Louis Henry Sullivan in 1895. *(Public domain)*

BELOW These terracotta panels featured on a number of Sullivan-designed buildings. *(Public domain)*

He began work as a junior architect in Philadelphia, but his employer was soon struggling to survive. A global economic crisis lasting from 1873 to 1879 cast a gloom over the architectural profession in the USA – except in Chicago, where there was plenty of work in the aftermath of the Great Fire of 1871. Let go by his first employer, Sullivan made his way to Chicago, where he soon found work as a draughtsman, for a certain William LeBaron Jenney, one of the pioneers of steel-framed buildings.

Sullivan soon moved on again, to Paris, where he studied at a prestigious art college for a year. Returning to Chicago he found work for a while as a designer of fresco stencils, then, in 1879, his opportunity to achieve greatness presented itself.

Pioneering Chicago architect Dankmar Adler gave Sullivan a job, and a year later took him into partnership. So began the firm of Adler & Sullivan, which soon gained fame designing theatres, working on projects across the USA. Their most famous creation was the extraordinary 1889 Auditorium Building in Chicago, which still stands.

This combines a 4,000-seat opera theatre, a 17-storey office tower, a ten-storey hotel and commercial storefronts at street level. Caisson technology was in its infancy, so with bedrock some 30m (98ft) down, the designers had no option but to try to construct stable foundations in Chicago's notorious soft blue clay.

A so-called floating raft of criss-crossed timbers and railroad sleepers encased in concrete formed a base for a series of rubble-and-stone pyramid-like piers that would support the loads for the 82m (270ft) structure above at key locations.

The team gave considerable thought to the differential settlement that would be caused by the taller tower, particularly given the fact it was planned to be built after the ten-storey section, meaning settlement would not only occur at different depths but also at different rates. Heavy iron and concrete blocks temporarily installed in the basement beneath the planned tower cleverly simulated the weight of the completed structure to help deal with this.

But design changes after the foundations were complete – not least changing the auditorium's walls from brick to the much heavier granite and limestone – resulted in

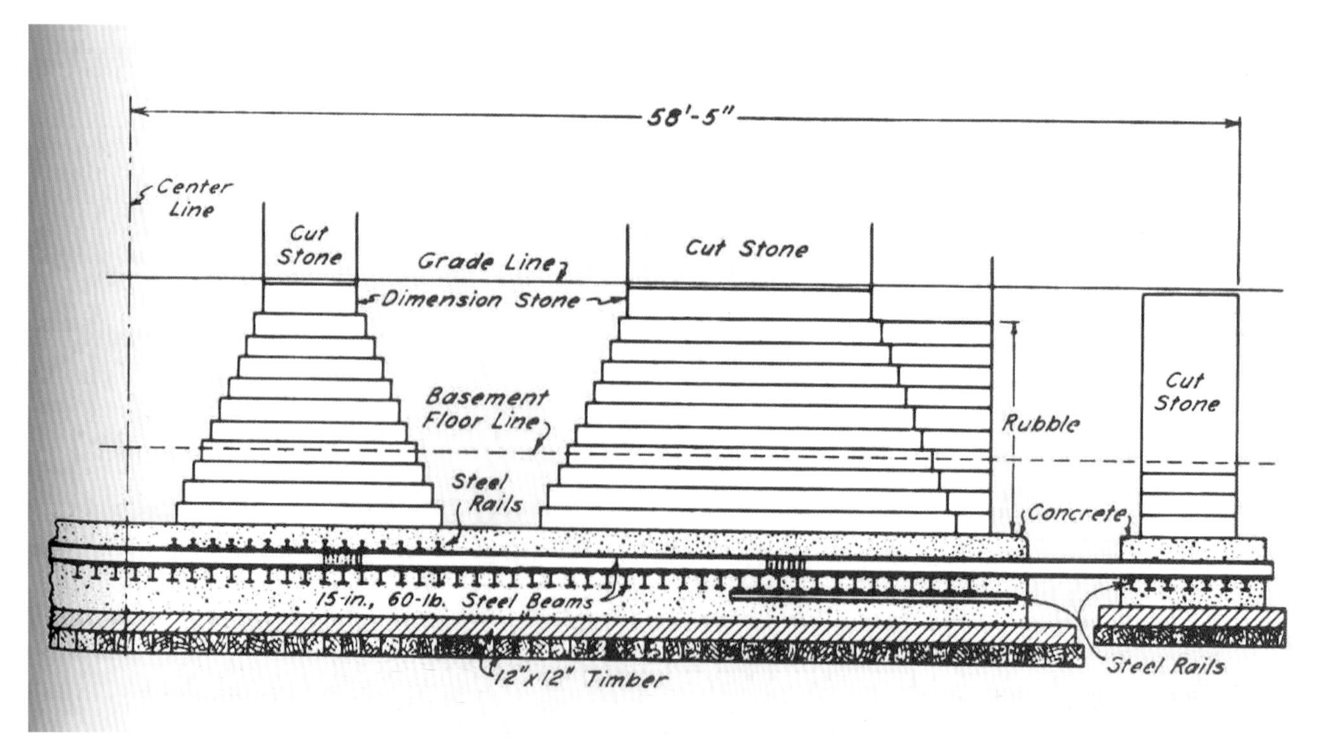

BELOW This foundation design for Chicago's 1889 Auditorium Building was ineffective and the building experienced extreme settlement of 750mm (2.5ft). *(Public domain)*

such extreme settlement – up to 750mm (2.5ft) – that later on stairs had to be added down from the sidewalk into the lobby that had once been level with the outside.

At the time this was the tallest building in Chicago and the largest building in the USA. Sullivan's career blossomed. He was credited with coining the seminal phrase 'form follows function', although Sullivan himself attributed this to the Ancient Roman engineer Marcus Vitruvius Pollio. Commissions for many types of building came the way of Adler & Sullivan, including the 13-storey Chicago Stock Exchange Building.

Originally the team planned to use another foundation system that had become popular in the city. Long timber piers, known as piles, were being driven down into the harder clays that lay below the less stable softer clay to support the new generation of taller buildings. Adler & Sullivan also planned to use piles for the Stock Exchange Building, that would ultimately be finished just a few years after the Auditorium in 1894.

But the method was not without controversy and the act of driving the piles deep into the ground was deemed responsible for causing damage to neighbouring properties, leading to legal challenges that threatened to stymie skyscraper projects.

And so, Adler & Sullivan brought in civil engineer General William Sooy Smith who had developed a new foundation system that seemed like it might just solve all their problems. This was to become known as the open well caisson foundation. The Stock Exchange proved to be the first full-scale use of the system and its success led to its widespread use on skyscrapers for decades to come.

To replace lines of driven timber piles, Sooy Smith's plan involved hand-digging eight wells down into the load-bearing rock some 30m (100ft) below the site of the Stock Exchange. Each well had a diameter less than 2m (6.6ft) and as the shaft progressed, workers stabilised it with vertical tongue-and-groove timber lagging kept in place by horizontal iron rings, forming sections similar in appearance to a wooden barrel.

Unfortunately, high water pressure and soft wet clays at around 15m (50ft) below the sidewalk in parts of downtown Chicago presented additional challenges for this method. Here, excavating the soil first and adding the lagging and ring caissons second was simply not an option – the shaft walls would buckle and water would flow in faster than it could be got rid of with buckets winched up to the surface.

Instead, the shaft was widened out above where this troublesome stratum was to be encountered (studies of the soil as it was excavated gave important clues as to when this was likely to occur).

Once the shaft was hand-dug to a diameter 150mm (6in) wider than the caisson section above, workers would sledgehammer pre-cut lengths of tongue-and-groove into the soil below to create an initial set of circular lagging in the ground. Then they would sledgehammer a second set of lagging inside the first at the same diameter as the regular caisson sections. The function of the first, wider, lagging was to help retain the weaker soils and provide some water cut-off for the inner shaft.

The second, inner, lagging could then be dug out, and the usual rings installed inside to stabilise the shaft and enable construction to advance toward the stable rock below. As the diggers reached the bottom of each deep well, they dug the base out wider still to create a bell bottom, known today as under-ream, giving an even higher-capacity foundation.

At full depth the shaft could then be concreted in from the base, with its now stable footing on top of the sturdier limestone bedrock, upwards to form the final caisson foundation. This allowed building work to begin above ground and, eventually, transfer of the Stock Exchange's loads through the piles down on to the bedrock.

Adler & Sullivan followed on with one other major project – the 12-storey Guaranty Building in Buffalo, New York, which still stands. Then in 1893 another economic panic struck, and the commissions began to dry up.

Soon after the partnership was dissolved and Sullivan's career went into long-term decline. Beset by chronic financial problems he turned to alcohol for relief. His marriage broke up and sadly he died alone in a Chicago hotel room in April 1924.

Chicago

Tribune Tower, 1925

The skyscraper with a Gothic crown

It all started as a publicity stunt for the influential *Chicago Tribune* newspaper. In 1922, to mark its 75th anniversary, it announced an international design competition for its new headquarters. The winner would receive what for the times was an astounding $50,000, and the *Tribune*, it hoped, would move into 'the most beautiful and distinctive office building in the world'.

In terms of drumming up publicity, the competition was a resounding success. There was interest among architects worldwide, and some very prestigious names submitted entries. Some bizarre designs arrived as well: one such example involved a tower topped by a gigantic sculpture of a Native American head.

Much to the surprise and dismay of the more innovative US architects, the winner was a derivative neo-Gothic design obviously inspired by the 1913 Woolworth Building in New York, then the tallest skyscraper in the world. Like the Woolworth Building, the 141m (463ft)-tall tower proposed by New York architects John Mead Howells and Raymond Hood was topped by an elaborate confection of ornate stone buttresses and pinnacles, all hand-carved.

This crown mimics the late Gothic Tour de Beurre of France's Rouen Cathedral. There are neo-Gothic decorative features throughout the exterior and interior, otherwise the structural design is conventional. The *Tribune*'s choice, coming as it did towards the end of the fashion for neo-Gothic towers, could hardly be said to be the most distinctive office building in the world – as for its beauty, opinions were mixed, then and now. It is, however, something of a must for modern-day tourists, and the newspaper is still based there.

Among the entrants to the original competition were some eminent foreign architects, including German luminary Walter

OPPOSITE **This won the design competition – but attracted few plaudits.** *(Chris6d)*

RIGHT **Houston's Gulf Building was inspired by Finnish architect Eliel Saarinen's second-place design.** *(i_am_jim)*

LEFT An Art Deco-style design was preferred for Gulf Oil's Pittsburgh headquarters. *(Derek Jensen)*

Gropius, founder of the influential Bauhaus School of modernist architecture. However, it was the design by Finnish architect Eliel Saarinen that received most plaudits from his peers, though in the end he had to be content with the $20,000 second prize.

Runner-up it might have been, and never actually built, but Saarinen's Art Deco design had a huge influence on 20th-century architecture. The next generation of skyscrapers, particularly in New York, was unashamedly Art Deco-inspired – controversial though it was at the time. Landmark towers such as the Chrysler Building and the Rockefeller Center are now accepted as architectural triumphs.

Houston, Texas, is home to what is generally regarded as the nearest realisation of the Saarinen entry. Originally known as the Gulf Building, now the JPMorgan Chase Building, the 131m (428ft) tower was opened in 1929. For more than 40 years the tower was topped with a distinctive rotating illuminated Gulf Oil sign.

Gulf Oil's main headquarters, however, was in the 177.4m (582ft) Gulf Tower in Pittsburgh, Pennsylvania, opened three years after the Gulf Building. Another Art Deco-style skyscraper, the building was topped with a stepped pyramid illuminated by neon lights that changed colour to provide a basic weather forecast visible for miles around. Although the tower still stands, the weather forecasts have been discontinued.

Empire State Building, 1931

The most famous skyscraper in the world

For over 40 years the 102-storey, 381m (1,250ft)-tall Empire State Building in Manhattan stood unchallenged as the tallest building on the planet. It became an icon, as much a symbol of New York City as the Statue of Liberty. Every year hundreds of thousands of visitors ascended to its two viewing platforms,

OPPOSITE This architectural icon was known around the globe. *(Sam Valadi)*

the highest at the 102nd floor – and a handful threw themselves off it.

In 1933 Hollywood made a star of the Empire State when King Kong roared defiance from its summit and clawed fighter planes from the sky. Yet for nearly 20 years the lighted windows in the fashionable Indiana Limestone Art Deco façade concealed an embarrassing secret – there were almost no tenants within.

What became known as the Great Depression is considered to have begun in 1929 with the stock market crash on Wall Street. This did nothing to dent the optimism of some property speculators in New York. Instead they seemed to become obsessed with a race to build ever higher than their rivals. The Woolworth Building had held the world record since 1913 but lost it to the 71-storey Bank of Manhattan Building when it opened in 1930. Other record-breakers were proposed, but only two projects actually went ahead in the face of the gathering economic storm clouds.

One such optimist was car manufacturer Walter Chrysler, founder of the giant Chrysler Corporation. Determined to build a record-breaker, Chrysler kept the planned height of his proposed skyscraper secret even after construction began.

Another optimistic car builder then entered the race. John Jakob Raskob, a former vice president of the equally gigantic General Motors, along with other wealthy backers, purchased the opulent but old-fashioned 36-year-old Waldorf-Astoria Hotel on 5th Avenue. An adjacent plot was also acquired, creating a 130 by 61m (427 by 200ft) site that could accommodate a major development. This was to be the site of the Empire State Building.

When proposals for the 282.6m (927ft) Bank of Manhattan Building were first made public in 1925, its planned final height was kept secret for a long time. Chrysler opted for the same strategy, leaving Raskob to respond as best he could. This competition was soon dubbed the 'Race to the Sky' by the local press. The Empire State team originally estimated that they would have to go to 80 storeys to top the Chrysler Building – but that kept growing. So five more floors were added to the Empire State design, which took it just 1.2m (3.9ft) above its rival.

Raskob was still not satisfied. He suspected that Chrysler might come up with a cunning plan once the Empire State was completed, such as smuggling a tall rod up through the Chrysler Building and raising it through the spire to claim back the record.

By this time the design for the Empire State had gone through many changes. Raskob's final instruction to architectural practice Shreve, Lamb & Harmon was to top the tower with a steel-clad tapering 'hat', a hollow spire with no internal floors between the 86th and 102nd floor levels. There would be an elevator linking the two, to take passengers to the foot of a series of steep ladders that would give access to a planned airship docking station at the equivalent of the 106th floor.

This took the planned overall height to 381m (1,250ft) – potentially a new world record. There were no signs of any challengers emerging in the near future, so Raskob and his backers could be quietly confident about retaining the crown for much longer than the Chrysler Building's 12-month reign.

Demolition of the Waldorf-Astoria began in October 1929. At the time it was the largest hotel in the world, so disposing of the debris was a major logistical challenge. In the end most of it had to be shipped out to sea and dumped. Then foundation construction could begin.

There was a theory that the tallest skyscrapers in Manhattan were located where the bedrock was closest to the surface. This seems to be flawed. Location was much more important to property speculators. Luckily, by the time of the 'Race to the Sky', the pneumatic caisson technique, as pioneered on the Manhattan Life Building in 1893, was well developed (*see* The three key developments, p.19), so economically viable skyscraper foundations were possible right across the island.

On 5th Avenue, a very desirable location, bedrock was 17m (56ft) down. No fewer than 600 men working two 12-hour shifts began excavating the Empire State's foundations in January 1930. So fast did they dig that the first steel column could be bolted to its concrete base just two months later.

This was the first of some 210 columns that made up the structural frame. Only 12 of these ran the full height of the tower: New York's

LEFT Despite this optimistic artist's impression, only one airship ever managed to dock with the Empire State Building. Turbulent wind conditions above the building led to the project's abandonment. *(Public domain)*

complex zoning laws dictated a major setback of the structure at the sixth-floor level, with the floors below this significantly larger than the upper floors.

Erection was rapid despite the logistical challenges. As is the norm today, all the steel elements were prefabricated offsite. In 1930, however, tower cranes with the capacity to lift girders up to 100 floors above street level were just a builder's fantasy. Instead, derricks – small, simple cranes – were installed one above the other as the steel frame headed skywards, finally creating a relay team that would pass a load up lift by lift until it reached its destination.

Once the steel sections were swung into position they would be held together with temporary bolts, then carefully aligned before a team of skilled riveters would make the permanent connection. Many of these teams would be Mohawks from Canada. (*see* The skywalkers of Manhattan, p.98).

Day by day the structure rose at breakneck speed. It was claimed that in one ten-day period more than 14 floors were erected. At peak there were almost 3,500 men working on the project on a single day. To match the workflow, more than 200 trucks arrived at the site every day, and this in the middle of Manhattan. Fortunately, traffic on 5th Avenue was far less congested than it is in the 21st century.

In September the steel structure was topped-out at the 86th floor and two months later the mooring mast was complete. Work on the interior continued until early the following year, with the new record-breaker officially opened on 1 May 1931.

New Yorkers could now marvel at what had been created in their midst. What seems to have attracted more attention than the sumptuous Art Deco details inside and out was the massive bank of elevators in the first-floor lobby. Faced with the challenge of moving the expected 15,000 occupants up and down the tower as quickly as possible, the designers came up with an innovative solution.

They added seven banks of high-speed Otis elevators, the first ever, a grand total of 58. None stopped at all floors. Bank F elevators only stopped between the 55th and 67th floors, for example. Coupled with the new elevator speed of around 6m/sec (20ft/sec), this ensured that there would be no lengthy queues building up in the lobby.

Crowds immediately flocked to the viewing platforms. By the 21st century, even though the Empire State had long since lost its crown, visitor numbers were measured in millions annually. Paying tenants, however, took longer to arrive.

Opening as it did in the midst of the Great Depression, which lasted the ten years until the outbreak of the Second World War in 1939, Raskob's dream soon collided with harsh reality. It was not until 1952, during the post-war boom, that the Empire State became even marginally profitable. Then in 1972 its record was overtaken by the original World Trade Center's 417m (1,368ft) North Tower. Nevertheless, its iconic status still remains unchallenged, and the lessons learned during its construction had a major influence on skyscraper design for many decades.

BELOW There were 58 Otis high-speed elevators in the new record-breaker. This is the sumptuous Art Deco entrance lobby. *(Fletcher6)*

Seagram Building, 1958

The height of luxury

In 1958 there opened on Manhattan's prestigious Park Avenue the most luxurious skyscraper ever built up to that time. No expense had been spared. More than 1,360t (1,500 US tons) of expensive bronze along with Italian marble and travertine limestone had been lavished on the 157m (515ft)-tall headquarters of Canadian distillers Joseph E. Seagram & Sons. Its height may have been relatively modest by Manhattan standards, but the Seagram Building was to become a major influence on skyscraper architecture for many decades.

It was a dream commission for prominent German-American architect Ludwig Mies van der Rohe (universally known as just Mies). Seagram gave him an unlimited budget. Unlike a typical commercial office development, there was no pressure to build quickly, to build cheaply or to cram in as many tenants as possible. Mies really had a clean sheet of paper. He was able to apply the architectural principles he had developed during his long and turbulent career to create what was immediately hailed as a masterpiece.

Born Maria Ludwig Michael Mies in 1886 in what was then the Kingdom of Prussia, Mies first started work in his father's monumental masonry workshop, but by the age of 22 he was employed in a Berlin architectural practice. He worked alongside such luminaries as Walter Gropius and Le Corbusier and was exposed to their progressive architectural theories. Along the way he changed his name to something more prestigious, adding the Dutch 'van der', because as a tradesman's son he had no right to the aristocratic 'von der'. Rohe was his mother's maiden name.

Before the First World War Mies established himself as a designer of traditional luxury homes for the wealthy and began a short-lived marriage that nevertheless produced three daughters in five years. He served with the German military during the war, returning to divorce his wife and resume his career designing neoclassical mansions. This was not enough. He began to develop a new modernistic style, and gained fame through a number of visionary projects,

LEFT One of the most influential skyscraper architects of all time, Mies van der Rohe's masterpiece is New York's Seagram Building. *(Public domain)*

including two glass-clad skyscraper designs that were never built.

However, by the 1930s the commissions were drying up, and Adolf Hitler had come to power. The futuristic designs of Mies and his associates were officially condemned as 'non-German'. In 1937 he was offered the post of head of architecture at the Illinois Institute of Technology in Chicago. With his situation in Berlin now precarious, Mies left for America, albeit with some reluctance.

The institute was a brand-new establishment, and Mies was given the responsibility for designing all the buildings on the campus. These stand to this day. With his reputation in the USA firmly established, Mies became an American citizen in 1944.

By the time the Seagram Building commission came along, Mies had a number of landmark buildings under his belt, most notably perhaps the twin 82.3m (270ft) 26-storey glass and steel residential towers on Chicago's Lake Shore Drive. Controversial at the time, these were later hailed as some of the first examples of the International style that came to be widely adopted in subsequent decades.

With New York land prices at such a prestige location as Park Avenue astronomical, the conventional wisdom was that every square metre of plot should be fully exploited. Commercial towers basically opened straight on to the sidewalk. With his unique design

freedom, Mies opted for a revolutionary alternative. The new Seagram Building would occupy less than half of the plot. It would be set back from Park Avenue, allowing for a pink granite-paved public plaza in front, complete with reflecting pools and marble benches.

In due course New York's zoning laws were amended to encourage developers to follow Mies's lead and create 'privately owned public spaces' as part of any new skyscraper proposals. The Seagram plaza has proved enduringly popular with the New York public.

At the heart of Mies's design philosophy was the mantra 'form follows function'. He believed that the building's structure should be visible, creating a 'more honest conversation' with the public than any applied decoration. As was the norm at the time, the main structure of the Seagram Building would be steel. New York's building codes required the steel to be protected against fire. In the days before the development of intumescent protective paints (*see* Dealing with disasters, p.110) this usually meant sheathing the steel in concrete.

This was unacceptable to Mies. Working with the assistance of architect Philip Johnson, he conceived a tinted glass curtain wall with non-structural bronze I-beams running vertically, in effect masquerading as structural mullions – window frames. Officially classed as 'topaz', the colour of the tinted glass was more often described as 'whisky-brown'.

This approach attracted some criticism at the time. Mies was also obsessed with the need to prevent the random use of window blinds – different blind heights at different windows 'disordered' the building's appearance, he insisted.

His partial solution was to specify blinds that could only be used in three positions: fully open, half open and fully closed. This attention to detail was mirrored in the luxurious interior, where bronze and marble featured in profusion. Included in the design was the exclusive Four Seasons restaurant, which was *the* place to eat in Manhattan for more than five decades.

Structural engineering firm Severud

OPPOSITE A seminal building that became an icon for generations of architects. *(Noroton)*

BELOW Creating a public plaza on half the building plot was a mould-breaker that created a template for many subsequent skyscrapers. *(Patrick Nouhailler)*

ABOVE Mies's last major project is commemorated on this stamp, issued shortly after his death. *(Public domain)*

Associates also took advantage of the financial freedoms on the project. This was to be the first tall building that used expensive high-strength friction-grip bolts to connect all the steel frame elements together (*see* Creating friction, below). The central core combined steel and reinforced concrete, there was an innovative vertical truss bracing system and the basic structural concept combined a braced frame with a moment frame (*see* Standing tall – the structural frame, p.82).

For the next ten years Mies continued to develop his 'skin and bones' architecture, and meanwhile also achieved international recognition as a furniture designer. His last major project was a new pavilion for the Berlin National Gallery, opened in 1968, that epitomises the Mies approach to architecture. A year later his lifelong heavy smoking habit finally took its toll, and he died of cancer in August 1969 at the age of 83.

His ashes were buried in Chicago's Graceland Cemetery, close to other famous local architects. The Seagram Building still stands as his memorial.

CREATING FRICTION

Conventional steel bolts are often used to connect two steel sections together. This allows loads and stresses in one section to be shared with the other. The boltholes in the steel have to be slightly larger in diameter than the shank of the bolt to allow the bolt to be inserted, and to make it easy to get both boltholes lined up.

Loads and stresses are transferred from one section to another through the shear resistance of the shank of the bolt (*see* diagram below). This limits how much load any bolt can carry, and the tolerances needed in practice can mean the connection could move slightly under load, potentially causing dangerous metal fatigue in the shank.

BELOW Quicker than welding, but too expensive for most developers. *(Anthea Carter)*

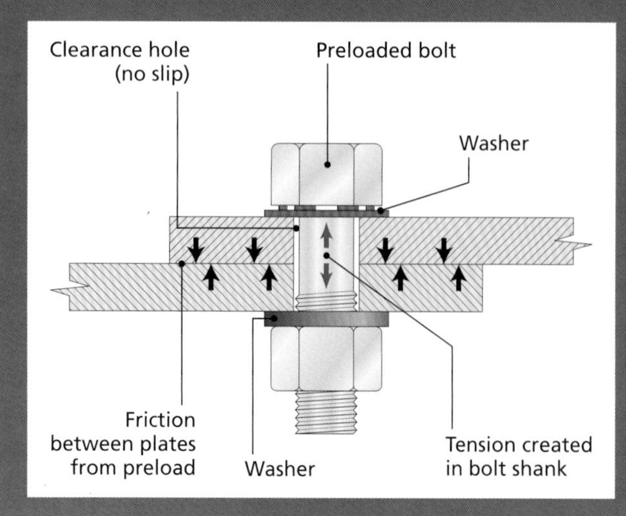

High-strength friction-grip (HSFG) bolts, as used on the Seagram Building, may look very similar to conventional bolts, but they work in a very different and more effective way.

A much higher grade of steel is used, which inevitably means HSFG bolts (also known as preloaded bolts) will be more expensive. This greater strength means the bolts can be tightened much further than conventional bolts, so the tension in the bolt shanks is very high.

Higher tension pulls the two sections very firmly together, so firmly, in fact, that the friction between them now acts as the load-transfer mechanism. The shear load on the bolt will be minimal, and there is little possibility of movement in service. In some applications the boltholes can be made significantly larger, speeding alignments and assembly.

HSFG bolts are mostly used on short- to medium-span bridges or similar steel structures, where the structure will naturally move more under load and where the avoidance of slip between structural elements is paramount. They are quicker to install than welding, but still not favoured by commercial property developers due to the extra expense.

Centre Point Tower, 1966

London's scandalous skyscraper

When it opened in 1966, the 34-storey, 117m (384ft)-tall Centre Point office tower in central London was the tallest habitable building in the city, topped only by the 177m (581ft) Post Office Tower nearby. Built primarily to facilitate microwave communications across London, the Post Office Tower (now called the BT Tower) was also famous for its revolving restaurant on the 34th floor, sadly closed in 1980 for security reasons. The Centre Point Tower became famous for less praiseworthy reasons. It lost its height record to the 183m (600ft) NatWest Tower (renamed Tower 42) in 1980, but long before that it had become a byword for the cynical financial manipulations of a certain type of property developer.

Record-breaking office towers have often struggled to attract tenants. Centre Point developer Harry Hyams turned prospective tenants away. He was hanging on for a single tenant who would rent the entire building on a ten-year or more lease and could afford to wait until the right one came along.

It was a long wait, 14 years in fact, during which the tower stood empty throughout a prolonged housing crisis in London. It was seen as 'an affront to the homeless' and triggered the formation of a homeless shelter in nearby Soho – which ironically christened itself 'Centrepoint'. The organisation was founded in 1969, and five years later a weekend occupation of the Centre Point Tower was staged by a group of housing campaigners.

There were also protests against the narrow pavements alongside the tower and its linked nine-storey block to the east. When crowded, there was a significant risk of pedestrians being forced out into the roadway. The area had the highest level of pedestrian injuries in central London, and the pavement layout remained unchanged for decades.

Hyams was unmoved. Even when prospective tenants made him lucrative offers to rent one or two floors, he refused to budge. Finally, in 1980, the tenant Hyams had been waiting for came along. The Confederation of British Industry rented the entire building and remained there for more than 33 years.

ABOVE This Central London skyscraper stood empty for many years – despite there being no shortage of potential tenants. *(Shutterstock)*

During its empty years, the tower's high visibility – at a time when there were few tall buildings in London – helped cement its image as a symbol of all that was wrong with the property market. Its prominence on the skyline was enhanced by its distinctive pale precast concrete façade, which incorporated expensive Portland stone aggregates from Dorset.

By the 21st century there was a surplus of modern office space in London. Centre Point could no longer compete with these upstarts. In 2015, under new owners, work started to convert the tower into residential accommodation.

RIGHT Originally known as the Sears Tower, the record-breaking Willis Tower took steel frames to their limit. *(Kelly Martin)*

Willis Tower, 1973

The birth of Chicago's record-breaking 'bundled tube'

Tasked with the design of a building to house the thousands of retail giant Sears Roebuck staff members then scattered in separate offices throughout the Chicago area, structural engineering firm Skidmore, Owings & Merrill (SOM) faced major challenges. The year was 1969, the tallest building in the world was still the 38-year-old Empire State Building and skyscraper design had hardly changed since its construction. If SOM was to provide the 280,000m^2 of office space originally specified by the client on a restricted city-centre plot at an economical cost, some clever engineering was needed.

To complicate matters, as design progressed, the client's optimism increased. More floor area was requested, so further floors had to be added to the provisional design. Soon the 100-floor mark was surpassed, and the proposed tower was growing ever closer to the limit imposed by the Federal Aviation Administration to protect flight paths into Chicago O'Hare Airport.

SOM's team, headed by Bangladesh-born structural engineer Fazlur Rahman Khan and architect Bruce Graham, knew that the classic steel skyscraper frame had already been pushed beyond economic limits on the Empire State project. They had to come up with a design for a record-breaking tower that would use significantly less steel than the Empire State, while at the same time yielding more usable space on every floor. A revolution in supertall skyscraper design was needed.

At 100 floors and more, wind forces represent the dominant loading on the structure. Khan started by going back to first principles. He saw that a supertall building was basically a vertical cantilever anchored to its foundations. As with a tree or a flagpole, wind-induced stresses would be greatest at ground level, least at the top. So a tapering tower would be potentially the most efficient structurally.

Khan had already realised that the best place for any structural elements designed to resist wind loads would be around the perimeter. His inspiration was bamboo, a very efficient structure in its own right. Locating the main structural elements at the perimeter also freed up lots of extra floor space and reduced the weight of the building dramatically. One downside was that as little as half the façade would be available for windows, marking a distinct shift away from earlier all-glass curtain wall designs.

This 'framed tube' concept was adopted with enthusiasm by many skyscraper designers. Khan had led the way in 1963 with the 43-storey DeWitt-Chestnut Apartment tower in Chicago. The original World Trade Center towers in New York were perhaps the most iconic example. But a tapered framed tube would be even more efficient – if an effective and practical way of achieving a floor plan that reduced with height could be devised.

A symmetrical taper such as that adopted much later for London's The Shard was never seen as a realistic option, given the level of cladding technology at the time. Instead, the design team went for a series of 'setbacks' rather like the classic New York style. The breakthrough came when Khan came up with the concept of the bundled tube.

He and Graham conceived a building made up of nine steel-framed tubes linked together in a three by three 'bundle' on a 69m (226ft) square footprint. As the building rose, individual tubes were terminated at three levels, leaving just two to reach the uppermost 108th floor (*see* diagram below).

No fewer than 13 floors in four groups at different levels housed all the essential environmental services and allowed internal trusses to be concealed. This design was significantly more efficient than the Empire State Building in terms of both material usage and usable floor space. A lighter structure meant smaller, more economic foundations as well. Furthermore, the design encouraged the use of extensive prefabrication, a key feature of many of Khan's tall-building designs.

At a stroke, architects were free from the tyranny of the basic framed tube design, which resulted in symmetrical monolithic buildings as epitomised by the Twin Towers. The opportunity to bundle the tubes together in many different ways was truly liberating. It also made possible the 'tuning' of a supertall's aerodynamics to minimise wind-induced sway.

One innovation, not required by the building codes of the day, was the installation of an automatic sprinkler system with around 40,000 sprinkler heads throughout the tower.

By the time what was then known as the Sears Tower opened in 1973 Sears Roebuck

BELOW Khan's concept of the bundled tube offered new freedoms to skyscraper designers.
(Anthea Carter)

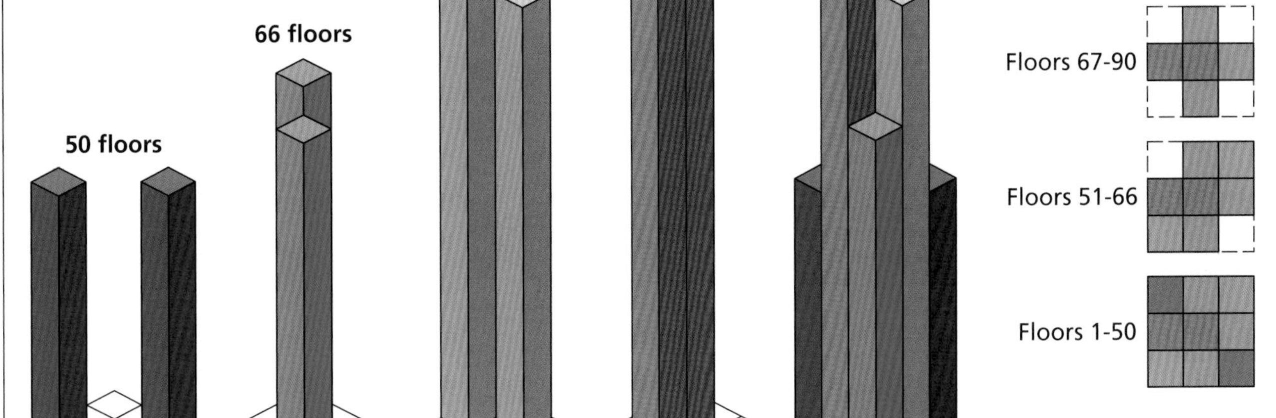

bosses were feeling a lot less optimistic about their future. New retailers were cutting into their market and the mass migration of all of Sears Roebuck's Chicago staff to the new tower never happened. Like most record-breaking skyscrapers, the Sears Tower struggled to make economic sense, standing half-empty for a decade or more. In 1992 even Sears Roebuck moved out.

More tenants departed post 9/11 amid fears that the Sears Tower, as the then world's tallest building, would be an obvious terrorist target. Indeed, a few years later a group was found guilty of plotting to destroy the tower, although the Deputy Director of the FBI at the time described the plot as being 'more aspirational than operational'.

Since 1994 the tower has changed hands several times. Finally, in 2009, it was renamed the Willis Tower after the London-based insurer Willis Group Holdings leased a section and purchased the naming rights.

From its opening until the present day, one feature of the Willis Tower that has always been successful is its observation deck on the 103rd floor – the highest in the USA – which attracts more than one million visitors a year. The tower still holds one record unlikely to be broken. It remains the tallest steel-framed building in the world. All subsequent record-breakers have either been concrete or composite structures.

One unusual feature in the Willis Tower's lobby is a wall-mounted sculpture of Khan. This may actually be a unique tribute. Other buildings sometimes have sculptures of their designers on show – but these are invariably architects. Structural engineers are usually the unsung heroes who turn architects' visions into reality. Khan, however, almost certainly had the highest profile of any structural engineer of the 20th century. During his professional career in the USA he was showered with awards and tributes, eventually being dubbed 'the Einstein of structural engineering'.

Khan was born in 1929 in what is now Bangladesh but was then Bengal, part of the British Empire. The son of a successful

BELOW Few structural engineers have been honoured in this way. *(Luthador)*

mathematics teacher, the young Khan attended high school before studying civil engineering at university in Calcutta (now Kolkata). After receiving his degree, he was awarded a Fulbright Scholarship, which gave him entry to the USA, plus access to the funds to support himself while he studied for further qualifications.

At the University of Illinois Khan eventually achieved two masters degrees and a PhD in structural engineering. He joined SOM in 1955, rising to partner status 11 years later. Despite never having seen any building higher than three storeys until he arrived in the USA aged 21, Khan soon became famous for his innovative skyscraper designs, of which the Willis Tower is the best known.

A pioneer of the use of computers as a design aid, Khan was also responsible for structures other than skyscrapers. Most notable of these were the enclosure of the McMath-Pierce solar telescope in Arizona and the Hajj Terminal at King Abdulaziz International Airport in Jeddah, Saudi Arabia. This is one of the largest in the world, capable of handling 80,000 passengers at the same time.

It was in Jeddah that Khan suffered a fatal heart attack at the age of 52. He was survived by his Austrian-born wife and a daughter.

Petronas Towers, 1998

Breaking the mould

When the 452m (1,483ft)-tall Petronas Towers complex in Kuala Lumpur, Malaysia, was completed in 1998, it became the world's tallest building, taking the crown from the Willis/Sears Tower in Chicago. This was a mould-breaking moment in many ways. Not only was it the first record-breaker to be located outside the USA, it was the first supertall building to use high-strength concrete for its main structure rather than steel; the first to link its twin towers with a sky bridge; and the first to adopt the architectural style known as Postmodern Islamic.

It also holds another record, one unlikely to be broken any time soon. Its foundations descend to unprecedented depths, due to its location on weak and fractured ground – probably the least suitable site for a supertall ever selected.

This was on the former home of the Selangor Turf Club, which moved away in 1988. Preliminary site investigations revealed up to 20m (66ft) of silty alluvium sitting on a variable layer of dense gravelly silt and clay. Below this was a very irregular stratum of heavily weathered limestone containing numerous cavities. At one side of the site the limestone rockhead was shallow enough to interfere with basement design. Worse still, it was discovered that halfway across the projected footprint of the towers the rockhead suddenly dipped into an underground valley of unknown depth.

This could have severely complicated foundation design and made differential settlement almost impossible to eliminate economically. The decision was made to shift the towers 60m (197ft) to the side, so they sat above the valley and therefore both towers would have basically the same ground conditions.

Some form of friction piling would be needed, with the rockhead so variable and so weathered that socketed piles would be out of the question. Extensive site testing established the length and number of piles needed to provide the load-bearing capacity. Unfortunately, this number of piles would put heavy loads into the clay immediately above the weak and fissured limestone.

LEFT Khan pioneered the use of computers for structural design. *(Public domain)*

RIGHT **Ground conditions below the Petronas Towers were extremely challenging, to say the least.** *(Anthea Carter)*

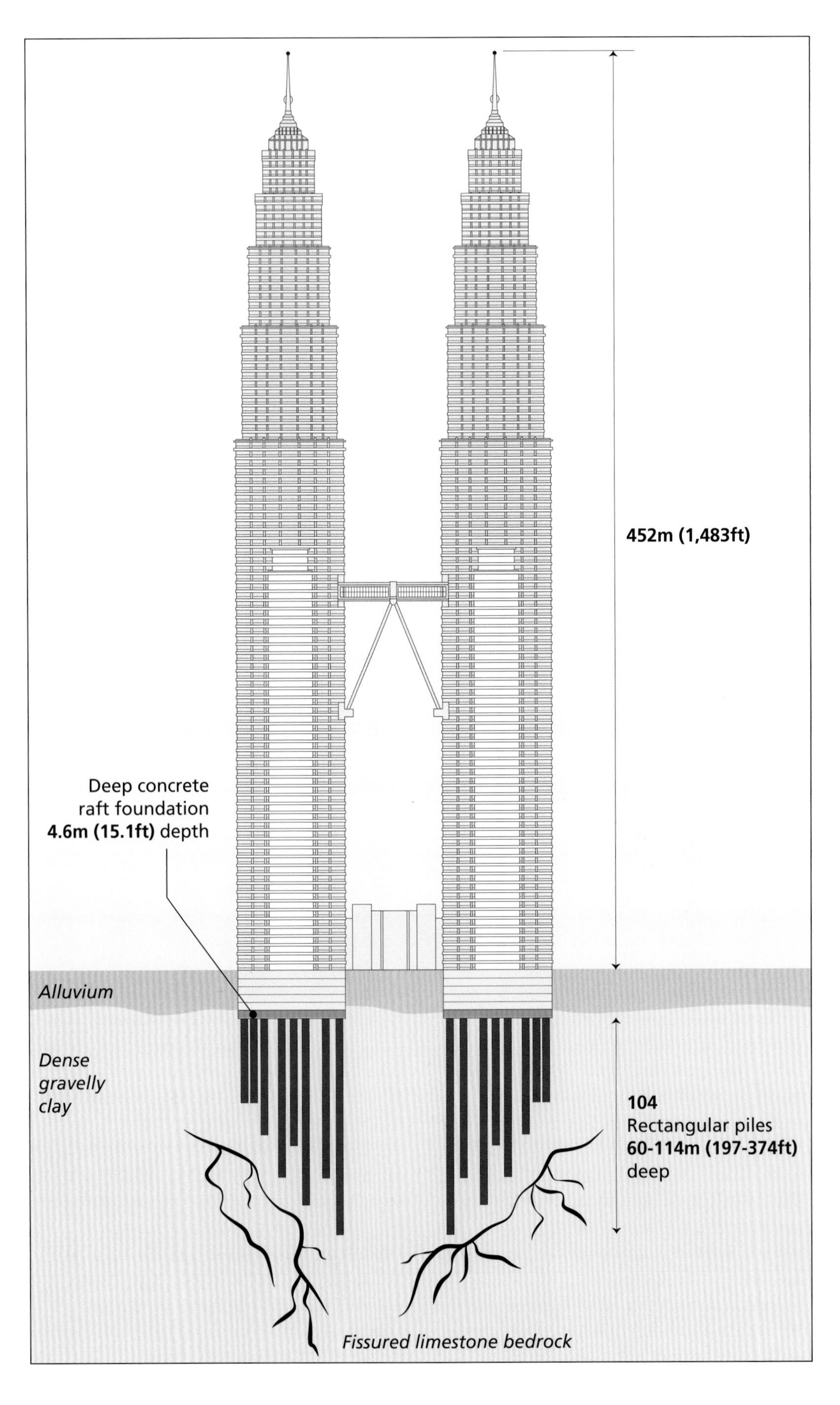

There were fears that the cavities could begin to collapse, causing massive settlement of the towers. A large-scale grouting operation had to be adopted, down to a depth of more than 160m (525ft). Hundreds of cubic metres of cement/water grout were needed to fill cavities that could be as large as 1m by 15m (3.3ft by 49ft).

To minimise the risk of differential settlement and to keep the buildings' maximum settlement to less than 13mm (0.5in), each barrette pile (*see* Digging deep, digging square, p.143) terminated approximately the same distance above the steeply sloping limestone rockhead. There were 160 piles in total; the deepest came in at 130m (426ft) – then, and now, a world record.

Another world record was set during the construction of the 4.6m (15.1ft)-thick concrete rafts linking each group of piles and forming the bases for the buildings' structures. More than 13,000m^3 of high-strength concrete were poured into each base in one 54-hour continuous operation. This record was only surpassed in 2007.

Structural engineer Thornton Tomasetti opted for a 'tube within a tube' structural design, akin in principle to the original World Trade Center towers, but with a much more robust concrete core. High-strength concrete was specified for the structural frame because there was existing expertise in its production in the country, and because importing steel would have been prohibitively expensive.

In a telling contrast to the severely rectangular and unadorned exterior of New York's Twin Towers, the complex geometrical forms of traditional Islamic architecture heavily influenced the architectural design of the Petronas Towers. Architect César Pelli based the floor plan on the two overlapping squares of the Rub el Hizb symbol, which is used in a number of flags and emblems in the Islamic world. This was developed into a practical floor plan with the addition of circular sectors that maximised office space and created what was then seen as a new language for supertall design.

RIGHT A new style of architecture for supertall towers was pioneered on the Petronas Towers. *(Public domain)*

RIGHT Hydrofraise machines can dig complex foundations. *(Anthea Carter)*

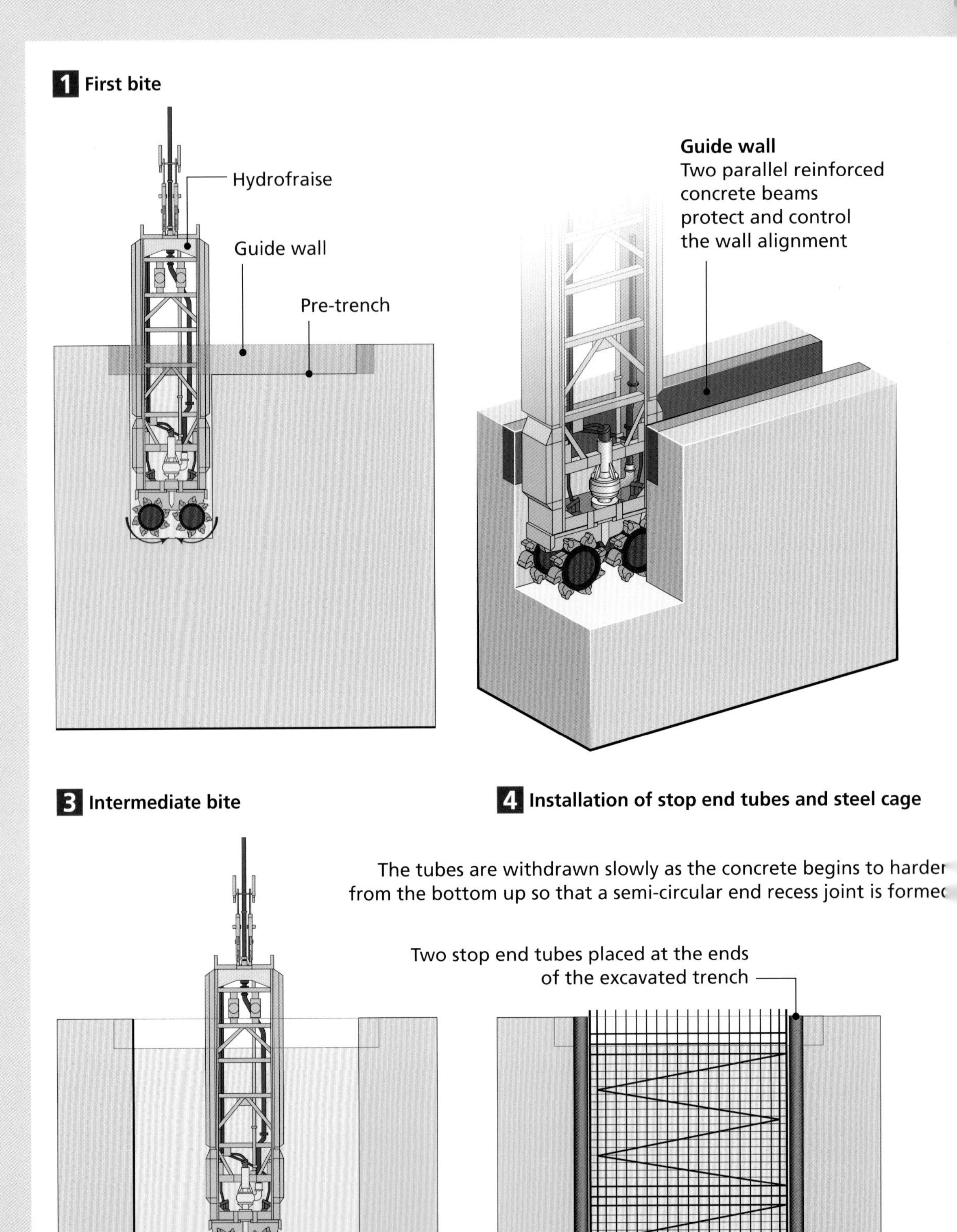

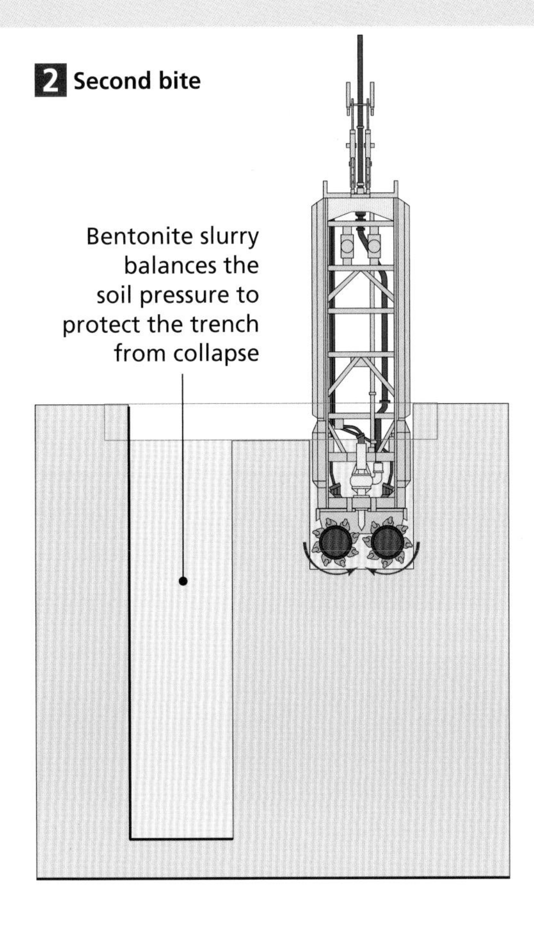

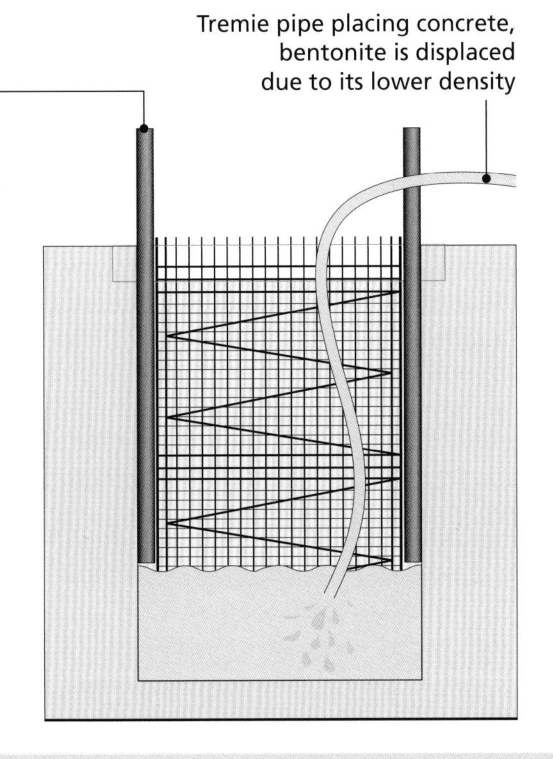

DIGGING DEEP, DIGGING SQUARE

Powerful rotary drilling machines can make possible deep, large-diameter, cast in situ concrete piles suitable for most applications. However, in extreme situations where the load-bearing capacity is totally dependent on the friction developed between the pile and the surrounding strata, a circular pile might have to be so deep to develop enough bearing capacity that it becomes impractical.

This was the case on the Petronas Towers project. The solution was the use of rectilinear barrette piles, which have a relatively higher surface area than the equivalent circular pile, and thus develop greater friction with the ground for any given length.

A number of different cross-sections have been adopted over the years, including X and T shapes. In the case of the Petronas Towers, a simple rectangular cross-section measuring 2.8m by 1.2m (9.2ft by 3.9ft) was deemed to be adequate. Barrette piles can be excavated using a simple crane grab, but for the extreme depths needed on the Petronas Towers project, the solution was the high-tech hydrofraise equipment.

In both cases the excavation is kept topped up with a bentonite slurry as it proceeds. The slurry is a gel-like mixture of an absorbent, sodium-rich clay and water that is dense enough to stabilise the walls of the excavation. A surface-mounted system constantly recirculates the slurry, cleaning it of all the excavated material.

At the business end of the hydrofraise is a pair of contra-rotating cutting heads. The verticality of the excavation is constantly monitored and controlled by an on-board computer.

Once the required depth has been achieved, the cutting head/crane grab is withdrawn. A steel reinforcement cage is inserted, a Tremie pipe is forced down through the slurry fill to the bottom of the pile and concrete is pumped in. This fills the pile from the bottom upwards, displacing the slurry as it does so.

ABOVE The Petronas sky bridge is a popular tourist attraction. Its role as an escape route during a mass evacuation is open to question, however. *(Public domain)*

As with all supertall skyscrapers, the Petronas Towers have to offer swift and reliable access to all their 88 floors. Each tower has a central core measuring 23 by 23m (76 by 76ft), housing 29 double-deck passenger elevators (*see* People movers, p.101) and six heavy-duty utility elevators.

A unique and eye-catching feature of the Petronas Towers is the two-storey 54.8m (180ft) long skybridge linking the towers at the 41st and 42nd floor levels. There are no rigid connections; the tower can accommodate any differential sway of the towers during high winds. Apart from its success as a tourist attraction, the sky bridge's planned function as an extra escape route during an emergency was found to be less than ideal in September 2001.

A bomb hoax the day after 9/11 triggered a simultaneous mass evacuation of both towers. There was chaos. The escape routes provided, including the skybridge, simply lacked the essential capacity. As a result, it was decided that the elevators could be used as well, and subsequent evacuation drills proved this approach to be successful.

After the Petronas Towers took the height record, the centre of supertall skyscraper construction shifted decisively away from the West to the booming economies of the Far and Near Easts. The first to top the Malaysian twins was the 508m (1,667ft) Taipei World Financial Centre (now called Taipei 101) in Taiwan, opened in 2004. Five years later both were dwarfed by the 828m (2,717ft) Burj Khalifa in Dubai. Even taller towers are planned at the time of writing, none of which are in the USA or Europe.

The Gherkin, 2004

Risen from the rubble

When the IRA detonated a 1t (1.1 US ton) truck bomb in the heart of the City of London on 10 April 1992, it could hardly have imagined that the end result would be the creation of one of the most iconic tall buildings in the UK – if not the world. Formally known as 30 St Mary Axe, universally known as The Gherkin, the building impresses by its unique architecture and structural engineering rather than sheer height. Taller towers have sprung up since it opened in 2004 – but everyone remembers The Gherkin.

St Mary Axe is a narrow street originally flanked by historic buildings, notably the Baltic Exchange, a global marketplace for shipping freight contracts. The 1992 truck bomb was parked right outside, while the IRA phoned a misleading warning that the bomb was intended for the Stock Exchange a short distance away. Three people died as a result, and 91 were injured. The façade of the Baltic Exchange was demolished and the rest of the building was badly damaged, as were adjacent buildings.

At first it was hoped that the original building could be resurrected, then, more realistically, that the façade could be reconstructed and incorporated into any redevelopment of the site. Those hopes faded when the extent of the destruction finally became clear. In the end, the remaining structures were dismantled and eventually sold off to Estonia. The plan was to incorporate them into a new commercial development. This never materialised, and the remains were still languishing in shipping containers at the time of writing.

This decision effectively created a prime development site. A number of proposals were put forward, including the 92-floor London Millennium Tower. This would have stood at 386m (1,266ft), making it the tallest building in Europe and the sixth tallest in the world. A unique double elliptical floor plan

RIGHT Universally known as the Gherkin, the Swiss Re headquarters immediately became a London landmark. *(Aurelien Guichard)*

was proposed, with a viewing platform 305m (1,000ft) above street level.

There were immediate objections to the plans, mostly centred on its extreme height and its inappropriate location. The death knell was sounded when Heathrow Airport to the west of London objected on the grounds that the supertall tower would disrupt flight paths from the east.

Media reaction was universally unfavourable. The *Guardian* newspaper immediately dubbed it the 'erotic gherkin', a tag that was soon taken up by most of the media. When the site was sold on to the Swiss Re global insurance giant, the name refused to die, and what eventually rose on the site inevitably became The Gherkin.

City planning authorities had already signalled that they would look favourably on an architecturally significant proposal, rather than a soulless rectangular block dedicated to providing the maximum net lettable floor area as cheaply as possible. When Swiss Re acquired the site for its UK headquarters it gave architects Foster + Partners almost the same free hand that the Seagram company had given Mies van der Rohe almost 50 years earlier – but not an unlimited budget. What was to rise at 30 St Mary Axe in London would become just as much a landmark as the bronze and glass masterpiece at 375 Park Avenue, Manhattan.

Its design was breathtaking – and revolutionary. London, especially the traditionally hidebound City, had never seen anything like it. It took care to minimise its visual and physical impact on the narrow streets around it while at the same time creating a unique

BELOW The Gherkin's entrance and plaza at night. *(Ian Muttoo)*

and unforgettable image that seemed to come straight from the fantasies of a science fiction writer. And if ever a building fitted perfectly into its nickname, it was The Gherkin.

Officially, of course, it was initially known as the Swiss Re Tower. Its floor plan was circular, tapering up from a diameter of 49m (161ft) at ground level to 56.5m (185ft) at the 16th floor and thence down to just 26.5m (87ft) at the highest floor, level 41. At a comparatively modest 180m (591ft) tall it sat well below any flight paths.

This aerodynamic shaping not only reduced wind loads on the tower, it all but eliminated unpleasant gusting at the foot of the tower. Visually, street-level impact is mitigated by a public piazza that gives access to an arcade of cafés and shops.

Turning Foster's vision into physical reality was down to structural engineer Arup. An external tubular steel triangulated diagrid forms the primary structure, allowing virtually unobstructed internal floors. Each floor plate is rotated 5° relative to the floor below. A double-skin glazing system forms the curtain wall, with a double-glazed outer leaf and a single-glazed inner.

Despite the building's curvaceous form, each of the 5,500 glazing panels, triangular or diamond-shaped, is completely flat. The only curved glass on the project is the lens-shaped cap at its apex. The glazing panels may vary only in size from floor to floor; the diagrid is a different matter.

Arup had to use 3D computer modelling to visualise the steelwork design. Every floor in the building is different – span, orientation and the angle at which each intersects with the diagrid varies from level to level. Connecting every element together in a cost-effective manner was the main challenge.

Some designers might have opted for elegant but very expensive cast steel connection nodes. Arup, however, devised a system of nodes fabricated from steel plates welded together at different angles. This was a very straightforward and practical solution.

One of the priorities for Foster was to create 'London's first environmental skyscraper'. The target was to use 50% less energy than a typical prestige office tower. One of the advantages of the curving curtain wall is that it maximises the penetration of natural daylighting deep into the floor plates, moderating the need for artificial lighting. This is further enhanced by the six light wells that spiral around the exterior.

These are formed by notches in each floor plate, with their function highlighted by grey-tinted glazing in the outer skin of the curtain walls. They also promote natural ventilation. Solar gain is minimised in the summer by blinds within the cladding system but encouraged in the winter. The need for heating and ventilation systems is significantly reduced as a result. There was also a focus on maximising the use of recycled or recyclable materials in the construction phase.

One of the reasons the City of London can be a challenging environment for property developers is the very real likelihood of uncovering the remains of millennia of continuous occupation buried not that far below modern-day ground level. Any such discovery can lead to significant delays to construction as archaeologists carefully conserve the findings. Thankfully, preliminary site investigations revealed no Roman villa or Tudor theatre, but they did come across the well-preserved lead coffin of a Roman matron, which is now on display in the Museum of London nearby.

One of the Gherkin's most popular features is located immediately below its apex. The bar here is available for hire, offering panoramic views of the City and the wider London area. Below it is a restaurant floor, and below that private dining rooms. This means there is no room at the highest level for machine rooms for the 22 lifts that service the lower floors. These only reach the 34th floor: a hydraulic 'push-from-below' lift bridges the gap to the 39th floor.

Since it opened in April 2004 the Gherkin has been garlanded with many awards. Its commercial history is not so glamorous. At the time of writing (2020) it is in the hands of a Brazilian group, the third owners in just 10 years. Valued at £700 million in 2014, it was then Britain's most expensive office building. It remains the most distinctive and best-loved skyscraper in the City, and most would agree 'it's worth every penny'.

RIGHT The Burj Khalifa in Dubai is the world's tallest building at 828m (2,717ft). *(Shutterstock)*

Burj Khalifa, 2010

The ultimate record-breaker

The Burj Khalifa's destiny as the tallest building in the world was clear from the outset.

Dubai wanted a new tower to trump the tallest of the tall and act as a dazzling beacon that would show the world how its new masterplan was backing its position as a burgeoning powerhouse.

During the late 20th century, the most populated city in the United Arab Emirates was experiencing enormous growth and economic success. It was evolving at a dramatic pace – leaving behind its heritage as a small trading port and emerging as the business and tourism epicentre of the Middle East's Gulf region.

The modern downtown Dubai needed a new centrepiece and developer Emaar Properties set its sights on a supertall tower. The idea was grand and encompassed the principles of compact urban living by creating a truly mixed-use space in a beachside setting. Not only would it include offices, retail space and residential units, but also a Giorgio Armani hotel. Bountiful green space with water features and generous pedestrian-friendly boulevards completed the grandeur at ground level.

Emaar hired Chicago-based Skidmore, Owings & Merrill (SOM), which set up a team led by the firm's notable architect, Adrian Smith, charged with developing a dramatic and innovative concept. What emerged in the final Burj Dubai tower design had echoes of SOM motifs on earlier projects and towers from its Chicago hometown. The distinctive form is notably reminiscent of the never-built mile-high tower that was the brainchild of famous Chicago architect Frank Lloyd Wright many decades earlier (*see* Super, mega and quirky – the battle to stand out from the crowd, p.61). SOM's Willis Tower project – also previously a world record-holder – lent its distinctive setbacks and tapering design as inspiration.

During its inception at the beginning of the 21st century, the Burj Dubai – as the Khalifa was then known – competed for the world's tallest title with the 508m (1,667ft) tall Taipei 101 in Taiwan.

With the benchmark set, the developer went

ABOVE The setbacks on Chicago's Willis Tower provided inspiration for the same feature on the Khalifa. *(Shutterstock)*

LEFT The distinctive profile is reminiscent of The Illinois, the never-built mile-high tower designed by Lloyd Wright half a century earlier. *(Shutterstock)*

ABOVE AND BELOW The Burj Khalifa's triple-lobed footprint is emblematic of the geometry of the regional desert flower, the hymenocallis. *(Shutterstock)*

about its plan. The actual intended height was kept a tightly held secret during construction for fear of other record-seekers dethroning it before or soon after its completion – the same tactics seen in action by New York building designers during the 1930s skyscraper boom. Work started on site in 2004 and on its opening in 2010, the Burj Khalifa's vast form was revealed as having topped-out at over 828m (2,717ft) – bypassing Taipei 101 by a staggering 320m (1,050ft), or the equivalent of a little more than the height of the UK's tallest building, The Shard. The 162-storey Khalifa and its more than 200m (700ft)-high spire secured the title, meeting all three major height criteria: the highest to its architectural top, the highest occupied floor and, of course, the highest from bottom to top.

Being the first tall building to go beyond the 600m (1,969ft) threshold meant that the Khalifa was the first skyscraper to be known as not just supertall, as defined by the Council on Tall Buildings and Urban Habitat, but the first to meet newly created criteria for being megatall.

Just as with some of the world's less tall but equally iconic skyscrapers, the project's promoters wanted it to be known for more than its height and have made much of how the building recognises and respects its location. Its form takes inspiration from the patterns seen in Islamic architecture and its triple-lobed footprint is emblematic of the geometry of the regional desert flower, the hymenocallis.

Built of reinforced concrete and clad in glass, the tower is in fact three elements formed around a buttressed central core. As it grows from its base, setbacks create an upwardly spiralling pattern and gradually reduce the building's mass as it reaches

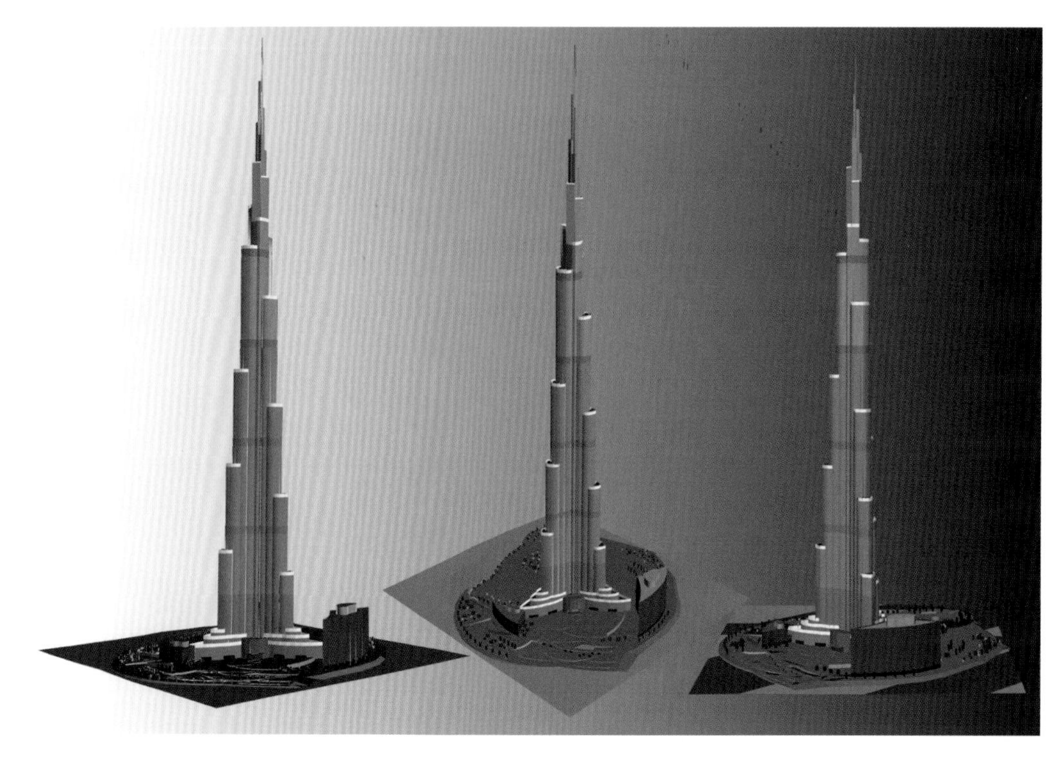

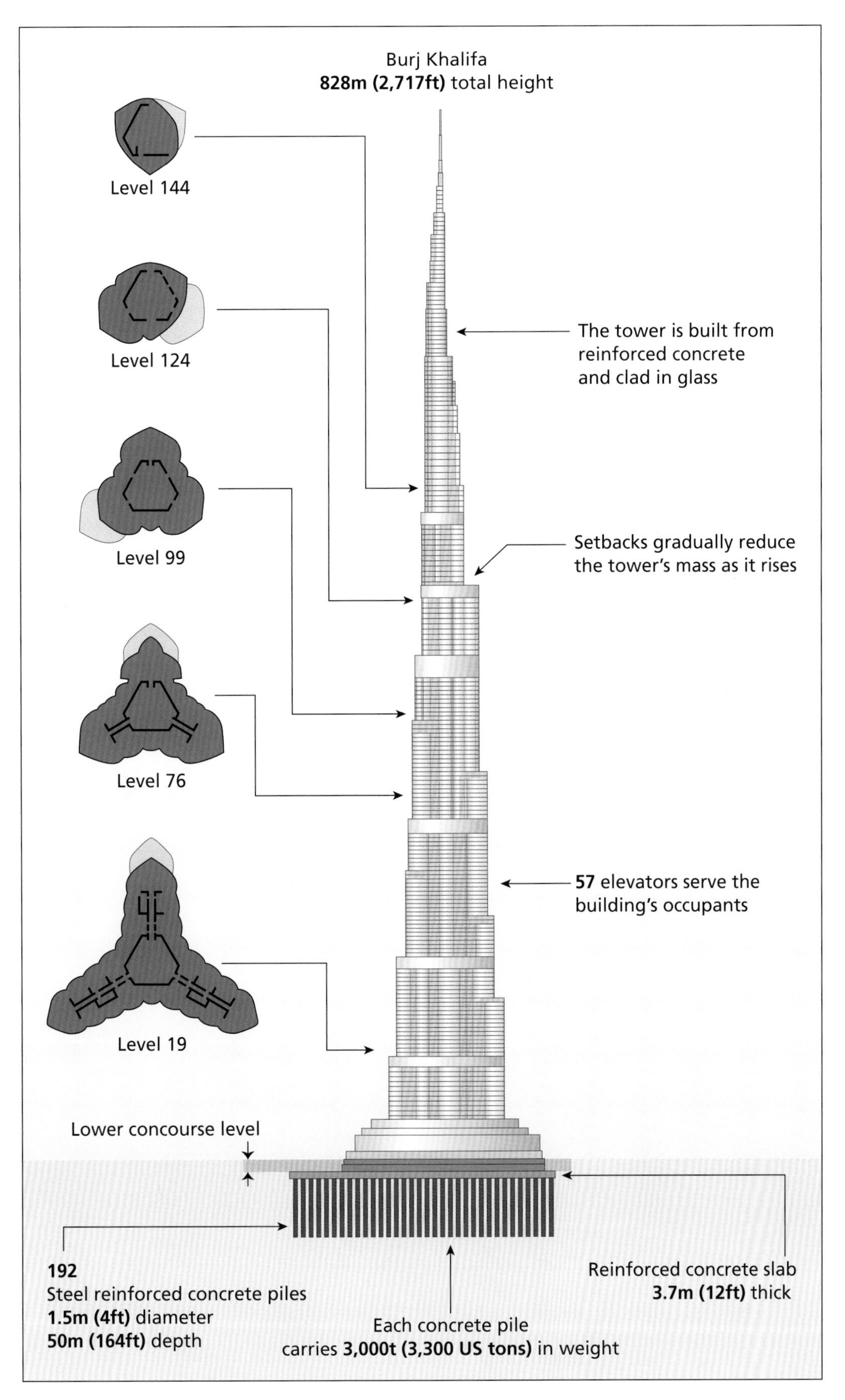

LEFT Built of reinforced concrete and clad in glass, as it grows from its base, setbacks create an upwardly spiralling pattern. *(Anthea Carter)*

its apex. At the pinnacle, the central core is revealed to form a spire.

The Y-shaped design of the floor plan is significant. Not only does it generate wide views of the Arabian Gulf for inhabitants, but it also performs an important structural function. Substantial wind-tunnel testing helped the team, which included engineering firm Hyder, to come up with the most efficient and stable design. Each lobe, or wing, comprises a high-performance concrete core and perimeter columns that buttresses the others via a hexagonal central core. Together this makes the tower very stiff torsionally.

Setbacks as the tower rises are achieved by aligning columns for the stepped level above the walls of the level below in order to smooth the loads. As well as forming the spiralling and tapering design, these setbacks help to effectively 'confuse the wind'. The idea being that the wind vortices that swirl around a building are disrupted before they build up and cause it to sway uncomfortably, thanks to setbacks that disrupt its otherwise smooth form.

Stability for the megatall tower below its lowest basement level comes from a 3.7m (12ft)-thick reinforced concrete slab. Beneath this is the real stabilising force in the form of 192 steel-reinforced concrete piles with a very large diameter of 1.5m (4ft) plunged to depths of around 50m (164ft). Because the insubstantial and loose soils across the site mean the bored concrete piles are socketed into weak rock at their base, their capacity to support the tower above is provided mainly by the skin friction between the pile concrete and surrounding soil.

The Khalifa's foundations are a fine example of the evolution in understanding those early engineering principles that Adler & Sullivan were grappling with well over 100 years earlier. A major feature of the pile specification considers how well the piles can cope with the settlement of the vast structure above – in this case they were designed to cope with up to

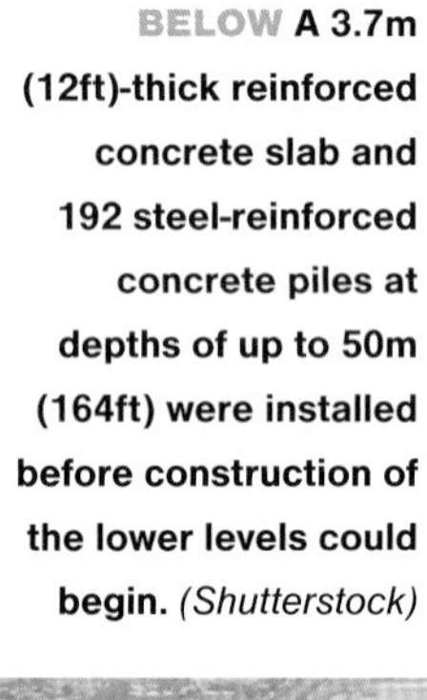

BELOW A 3.7m (12ft)-thick reinforced concrete slab and 192 steel-reinforced concrete piles at depths of up to 50m (164ft) were installed before construction of the lower levels could begin. *(Shutterstock)*

80mm (3.15in) of movement. Each individual concrete pile beneath the Khalifa carries 3,000t (3,300 US tons) in weight and the total combined concrete and steel foundation weighs in at an extraordinary 110,000t (121,254 US tons) – coincidentally the same figure as Adler & Sullivan's Auditorium Building.

Serving the inhabitants of the world's tallest building are no fewer than 57 elevators and 8 escalators. Some are programmed to be used in the event of an emergency, in a trend that has seen tall-building evacuation plans move on from relying on the use of stairs alone. Observation decks for visitors are positioned at levels 124, 125 and 148, and elevators serving these attractions contain double-deck cabins, each with capacity for up to 14 people travelling at 10m/sec (33ft/sec).

Shrouding the vast structure is a curtain wall said to cover an area equivalent to 17 football (soccer) fields. This is made of aluminium and textured stainless-steel spandrel panels and stainless steel vertical tubular shaped fins, which along with reflective glazing, are designed to withstand the city's harsh summer heat.

ABOVE **The staggering height means much of the building is often shrouded in fog to the extent that building owners promote the possibility of visitors dining in the clouds.** *(Shutterstock)*

BELOW **Observation decks are positioned at levels 124, 125 and 148.** *(Shutterstock)*

In spite of, or perhaps because of, the challenging desert location, the tower still strives towards sustainable design and makes use of sky-sourced ventilation. This means that cooler air and the relatively lower humidity in the air at the top of the building is exploited and drawn in to the building, requiring less energy for ventilation, air conditioning and dehumidification.

Added to this is one of the largest condensate recovery systems in the world, where air conditioning condensation is collected instead of being allowed to be discharged as wastewater and is reused for irrigating the tower's landscaping.

Some 12,000 construction workers were needed to build the tower between January 2004 and late 2009 and the building officially opened to great fanfare in January 2010. The Burj Khalifa remains the tallest building record holder, though attempts to knock it off the top spot by the 1km (3,281ft)-tall Jeddah Tower in Saudi Arabia are under way, but construction was suspended at the time of writing. The Khalifa retains its title for now and continues to draw the crowds.

OPPOSITE **A curtain wall the size of 17 football (soccer) fields envelops the structure.** *(Shutterstock)*

BELOW **Some 12,000 construction workers were needed to build the tower, which opened to the public in January 2010.** *(Shutterstock)*

Chapter Five

The future

And what of the future? More skyscrapers will be built, some even taller than anything proposed today. New technology will make construction quicker and cheaper, the push for smaller carbon footprints will continue. We might even see radical new materials replacing traditional steel and concrete. Finally, we look back at the World Trade Center disaster, and what can be learned from it.

OPPOSITE If built, the 300m (984ft) timber Oakwood Tower would be the tallest timber building in the world and the second tallest skyscraper in London after The Shard. *(With thanks to PLP Architecture, Smith and Wallwork and Cambridge University)*

Wild ambitions

Since their inception, skyscrapers have kept pace with the changing needs and wants of developers and occupants alike, and their popularity looks set to rise. In recent times, the record for the number and height of skyscraper completions has been beaten year after year. And no matter what the concerns over security, climate, visual and social impacts, the appetite for building tall seems greater than ever.

The fact of the matter is that the tall building remains an efficient way to create spaces in which people can live and work as urban density and populations in cities and towns continues to increase. The number of 200m (656ft)-plus buildings typically well exceeds over 100 in number every year.

Amid these trends, one of the enduring questions is how high can we really go, or even, is there such a thing as too high?

If and when the Jeddah Tower is finally completed, the 1km (3,281ft) height mark should have been passed at last. Proposals for a mile-high hypertall structure are sure to follow. This would soar 1.6km into the sky and would see the realisation of famous architect

BELOW Some of the most notable tall-building record-holders over the past century (left to right) Burj Khalifa, One World Trade Center, Petronas Towers, Empire State Building, The Shard. *(Shutterstock)*

Frank Lloyd Wright's 1956 vision for 'The Illinois' (*see* Super, mega and quirky – the battle to stand out from the crowd, p.61). There are those who dream of even taller towers, true cities in the sky – but just what are the limits on building so tall?

First of all, the ground conditions must be favourable. Solid, unweathered rock would be the first choice. The enormous weight of a hypertall tower is just one factor, but equally important are the overturning forces generated by high winds. The first proposals for the Jeddah Tower were for a mile-high tower, but the ground at the proposed site was simply too weak to support such a structure without gigantic foundations.

A tapered solid-steel column could actually reach a height of more than 5km (3.1 miles) before it would collapse under its own weight. Similarly, a tapered normal-strength concrete column could achieve nearly 4km (2.5 miles). The USA's National Aeronautics and Space Administration (NASA) has even calculated that a solid column of carbon fibre-reinforced epoxy resin could actually stretch up as far as an incredible 114km (71 miles).

A structure on this scale would essentially be a forest or bundle of columns without floors. Wind and seismic loads would have to be considered as well, along with potential sway. Add in the need to be able to transport occupants to all floor levels in reasonable time, and the likely height limits for a real building are much lower.

Assuming a buttressed core design similar to the Burj Khalifa or the Jeddah Tower, a 100% normal-strength reinforced concrete tower could theoretically top 1km (3,281ft). An all-steel design could go significantly higher – 3.2km (10,500ft or thereabouts) – almost a 'two-mile-high' building, in fact. A hybrid steel/concrete tower would be somewhere in between.

All this assumes an aerodynamically efficient tapered profile. An untapered 'prismatic' tower will be limited to a maximum of two-thirds of the heights suggested above, and probably less, simply because of increased wind resistance.

It is almost certain that any realistic hypertall tower would be a mixed-use building. Apart from those floors devoted to offices, hotel accommodation and residential, many floors would have to be given over to all the support facilities such cities in the sky would need. Occupants of the 150th floor would not be particularly keen on descending all the way to street level to do their shopping, visit the dentist or have their hair cut, for instance. Moving occupants vertically and horizontally – or even diagonally – could be facilitated by the adoption of 'ropeless elevators' (*see* People movers, p.101), powered by linear induction motors. These would help to keep the size of the service core down to economic levels.

Finding the ideal site for a 3km (2-mile)-plus tower would be difficult. It would need ideal ground conditions, proximity to a transport nexus and enough potential occupants to make it a sound economic proposition. On the other hand, there will always be dreamers who might convince national or local governments that having a 'two-mile-high' building on their territory would confer benefits that would outweigh the need to balance the books. Hypertall towers will continue to rise – and maybe one day the first carbon-fibre example will reach up some 5km into the sky.

Semi-serious designs have emerged for enormous pyramid structures essentially containing mega-cities within them. Tokyo and San Francisco are among the locations being touted as potential future development sites for such a concept. Talk is of futuristic lightweight structural frames that could accommodate one million people with individual skyscrapers suspended from within the frame to create living, work, retail and entertainment spaces. Accelerated walkways, personal rapid transit systems and superfast elevators (vertical or inclined) with 10-minute transit times to their apex might connect the inhabitants.

The materials science to create these nearly 2km (1.2 mile)-diameter base and 500-plus storey structures is again the most obvious limitation to them becoming a reality any time soon. Advanced cable systems, or lightweight carbon nanotubes, still being researched, are likely to be key to unlocking serious discussion as to whether one could ever be built.

In stark contrast to these vast and wide structures, there is a new form of skyscraper that has really sparked a trend. And that is the super slim, luxury residential towers springing up in the chock-a-block environs of New York City.

The extraordinary feature of these new-style supertall towers is less their record-breaking height but more their pencil-like appearance. These space-maximising supertall structures require little land take, and are a response to the need for luxury apartments in the limited prime real estate available that provide for the expansive city views Manhattan is so famous for.

Flanking Central Park is an obvious hotspot for many developers of these slender skyscrapers, but other parts of the island are drawing in speculators too, given that if they can stretch these monoliths to the right heights, the widest views will be afforded. In under a decade the city has seen the completion or near completion of around 18 towers with this strikingly elegant type of profile that appear to jut out much higher into the sky than regular-width skyscrapers.

It is not simply a matter of personal judgement that makes these slimmer-than-usual towers defined as such. 'Slenderness' is actually a technical engineering definition. Between a 1:10 and 1:12 ratio of the width of the building's base to height, is the minimum structural engineers consider defining tall buildings as slender.

To illustrate, let us take a more conventional tall building design for comparison. At its completion in 1972 the World Trade Center North Tower took the title of the world's tallest building. But with a height of 417m (1,368ft) and large 63.7m by 63.7m (209ft by 209ft)-square floor plate, its width-to-height ratio was less than 1:7, which is in stark contrast to the recently completed 432 Park Avenue tower, whose 28m by 28m (93ft by 93ft)-square base and height of 426m (1,397ft), generates a slenderness ratio of 1:15.

While the monolithic look of some of these slender, and the more familiar wider generic, skyscraper towers is an all-too-familiar sight,

BELOW A new trend is emerging for supertall and slender skyscrapers, particularly in New York. *(Shutterstock)*

RIGHT 432 Park Avenue in Manhattan has a 28m by 28m (93ft by 93ft)-square base and a height of 426m (1,397ft), which makes for an especially slender profile. *(Shutterstock)*

the future design of skyscrapers will be judged much more harshly if not enough attention is given to the experience of inhabiting them, rather than simply how they look on the skyline.

Managing the effects of wind on these supertall and super slender structures is an even more pressing issue for engineers than with more regular-proportioned skyscrapers. If unaddressed, wind vortices would induce unbearable amounts of sway for occupants (*see* Standing tall – the structural frame, p.82). In response, wind-tunnel testing is used as a default to see how their structural features and façades perform.

ABOVE Vegetation on the outside as well as inside of tall buildings can offer environmental benefits. *(Shutterstock)*

OPPOSITE As urbanisation continues to draw more people to live and work in towns and cities, the trend for building tall looks set to continue. *(Shutterstock)*

In addition, designers are featuring jutting-out ledges and breaks and airflow gaps in these structures to disrupt the wind and reduce its negative effects. Mass dampers that counteract sway are also highly useful in accomplishing this.

At 432 Park Avenue, wind-tunnel testing of the early design revealed that the structure would suffer from significant vortex shedding, even in relatively low winds, causing the building to vibrate unacceptably. The design was altered to create gaps for the air to flow through the building at regular points by simply removing glazing from the mechanical floor levels. The building also houses two concrete tuned mass dampers on the 84th floor, weighing 1,300t (1,433 US tons) and supported by cables and hydraulic cylinders to counter the residual movement and slow it to manageable levels.

Increasingly, good tall-building design considers not just a structure's ability to withstand the forces of nature, but how it fits in with the natural environment too. They should be considered not as a single entity but rather as stacked communities, recognising the fact that the experience of those inhabiting the lower floors can be mightily different from those on the upper floors of a supertall or megatall building. To take just one example, the external air temperature outside the world's tallest tower, the Burj Khalifa, is several degrees cooler at the top than at the base.

Shared spaces in these stacked communities are likely to have to more carefully consider widespread landscaping throughout the internal spaces and how they might encourage social activities to take place. Vegetation not only inside but also on the external face of these buildings can aid environmental qualities, add shading and improve air quality within. Varying the form and how it relates to this changing environment throughout its levels will be critical in ensuring a successful and popular development.

Concentrating greater numbers of people high up in these larger-than-ever skyscrapers means making city conveniences more readily available. Amenities such as parks, schools, doctors' surgeries and other public offerings will need to be lifted up into these buildings too.

Much innovation is beginning to be forthcoming in modern buildings. Now ensuring that buildings can be constructed and operate within an increasingly carbon net zero-focused world will require an even greater effort.

Offsite building and the resurrection of prefabrication

Skyscrapers look set to stay in vogue in the future. Experts predict that population growth will continue and that the rise will be felt most alongside the effects of urbanisation, where increasing numbers of people choose to live and work in towns and cities. Creating more space in those already densely occupied places will more than likely mean building more and more high-rise towers to accommodate them.

What that also means is that developers will face continued pressure to come up with ever more safe, efficient and economic ways of building skyscrapers, with a focus on making them more technologically advanced and less draining on precious natural resources.

RIGHT **Building developers, engineers and builders are seeking ways to take risk out of skyscraper construction by using manufacturing techniques and recreating factory-like conditions on site as far as possible.** *(Shutterstock)*

Advancing technical know-how and maturing technology are expected to provide some of the solutions.

There are some exciting improvements in construction materials on the horizon that may yet disrupt current methods. The extraordinarily strong yet light characteristics of wonder material graphene could shift the current reliance on the resource-draining and carbon-intensive traditional concrete and steel in use in skyscraper construction.

Other gains can be made not so much from what they are built with but more from how they are built. Engineering is continually evolving the tools needed to build in smarter and faster ways.

Offsite building methods are becoming more and more useful. Terminology covering this concept varies from place to place (modern methods of construction/prefabricated, prefinished volumetric construction/permanent modular building), but the theory is the same: take as much work away from the often-harsh working conditions on site and instead fabricate as much of a building's components in factory-like conditions.

BELOW **Modularised components that can be quality controlled and mass-produced are seen as offering great cost, materials and safety benefits.** *(Shutterstock)*

Increasingly, manufacturers are treating tall buildings as engineered components that have all the necessary systems built in.

Building the structural core and frame, adding floors, a curtain wall and then fitting it out with all the heating, cooling and ventilation required is being eschewed in favour of prefabricated multi-layered cladding systems that embed internal wall finishes, including electrical wall sockets. Air conditioning, heating and ventilation, security, fire protection, lighting, energy management, water and elevator systems with software and hardware that manage them can be integrated and built in to the prefabrication process.

The same approach is being used on structural elements, flooring units and even whole rooms. Bathroom pods of increasingly high quality are being built with the bath, shower, toilet and plumbing ready to be slotted into place and plugged into the building's services.

This is not an entirely modern concept. Prefabrication is a well-known term that has associations with everything from simple precast concrete panels to the often derided million-plus homes that helped solve the UK's housing crisis in the aftermath of the Second World War.

But this time around, with the aid of technology, ambitious developers are hoping to shake off any negative past connotations by proving that high-quality, multi-faceted elements can be built to last, be easy to maintain and – even better – can be readily

Around 85% of The Leadenhall skyscraper's costs were spent on manufacturing components offsite. *(Shutterstock)*

ABOVE Detailed augmented and virtual reality versions of building designs are making previously hard-to-predict construction challenges apparent at an earlier stage, at which point they are more easily remedied. *(Shutterstock)*

transported to site and slotted together to form large-scale structures in super-quick time.

Less of the pouring of wet concrete on site means less reliance on bespoke formwork and fewer people working in weather-exposed, compact and sometimes unsafe construction sites.

With quality-controlled factory conditions there is also less waste and all the elements can be tested before being transported to site.

Completed in 2014, the 224m (735ft)-tall London skyscraper referred to as the Cheesegrater, or more formally The Leadenhall, is proud of its offsite credentials.

The building's distinctive and very visible yellow north-facing core houses 20 passenger lifts and was formed from 138 table-shaped components, all manufactured offsite and complete with services and structural flooring. Each unit weighed 35t (38.5 US tons) and was able to be hoisted into place within two hours of arrival.

All of the building's concrete floors were precast. Four cranes, two of them custom-built for the project, did the job of hauling the giant components into position. This was vital because these complex and large units need careful handling. Great care and expertise are required to ensure load handling is correctly identified – particularly given such units are bespoke and numbered so as to be designed to fit very specific parts of the building. If damaged in transit or on site they can be extremely costly to be remade and will likely cause unwanted delays.

As much as 85% of The Leadenhall's costs were spent on offsite manufacture – a huge proportion for a skyscraper and perhaps an indication that more and more will be built this way.

Advances in computer software and new technology hardware is helping to make this possible. Fully embracing digital tools means that skyscrapers and major construction projects are being built twice, so to speak – once in a virtual environment and once in the physical world. The point of this is to find out earlier if there are any issues that might normally be uncovered during construction. Changing designs and plans at a later stage when on site can be much more costly and more complicated to resolve.

Incredibly detailed augmented or virtual reality versions of building projects can reveal potential design clashes that make construction so inherently unpredictable at an earlier stage. While 2D technical drawings and human engineering expertise remain vital, enhanced digital versions of drawings enable faster communication between the many people involved in creating the building and is referred to as building information modelling (BIM).

There are a variety of levels of BIM now, from the more basic digital versions of traditional drawings through to full virtual reality-enabled models that hold not just visual cues to recreate the feel of a building if you are wearing the right headset and equipment, but also all the technical, logistical and operational data too. Extraordinary details including intricate measurements of every element, accurate and highly specific building materials information, construction sequencing, all the financial costs and maintenance regimes can all be held within the software model.

These impressive virtual versions of the physical world are being tagged as digital twins.

What this means is that there is now an increasingly important role for computer programmers and data scientists in building the world's tallest skyscrapers. All of it being done by computers and robots is probably still far from reality. And still the widespread rollout of technology across construction depends on whether the people paying for them are willing to invest in often higher up-front costs that are so often associated with new ways of doing things.

The rise of the jumping factory

Developments in construction methods are allowing for the more rapid and efficient building of skyscrapers, primarily by taking critical building processes away from the risky construction site and into factories. Today, however, there is still a need for work to eventually move on site to enable all the elements of a building to be brought together. And so developers and builders seek new and innovative ways to make this phase of work safer and more efficient.

The key to modern ways of thinking around this is to try to emulate controlled factory working conditions and apply them to on-site construction activities as far as possible.

The recent emergence of rising factory operation is one prime example of this way of working. In essence a multi-level workspace is formed around the building and can be climbed – or jumped – up the structure as it rises from the ground, revealing finished floors with internal rooms built entirely as the rest of the structure continues to be formed above.

Huge gantry cranes capable of lifting in vast wall units and large floor slabs are housed within these tented factory enclaves that span up to ten storeys high. Within this

BELOW The concept of the rising factory. *(Anthea Carter)*

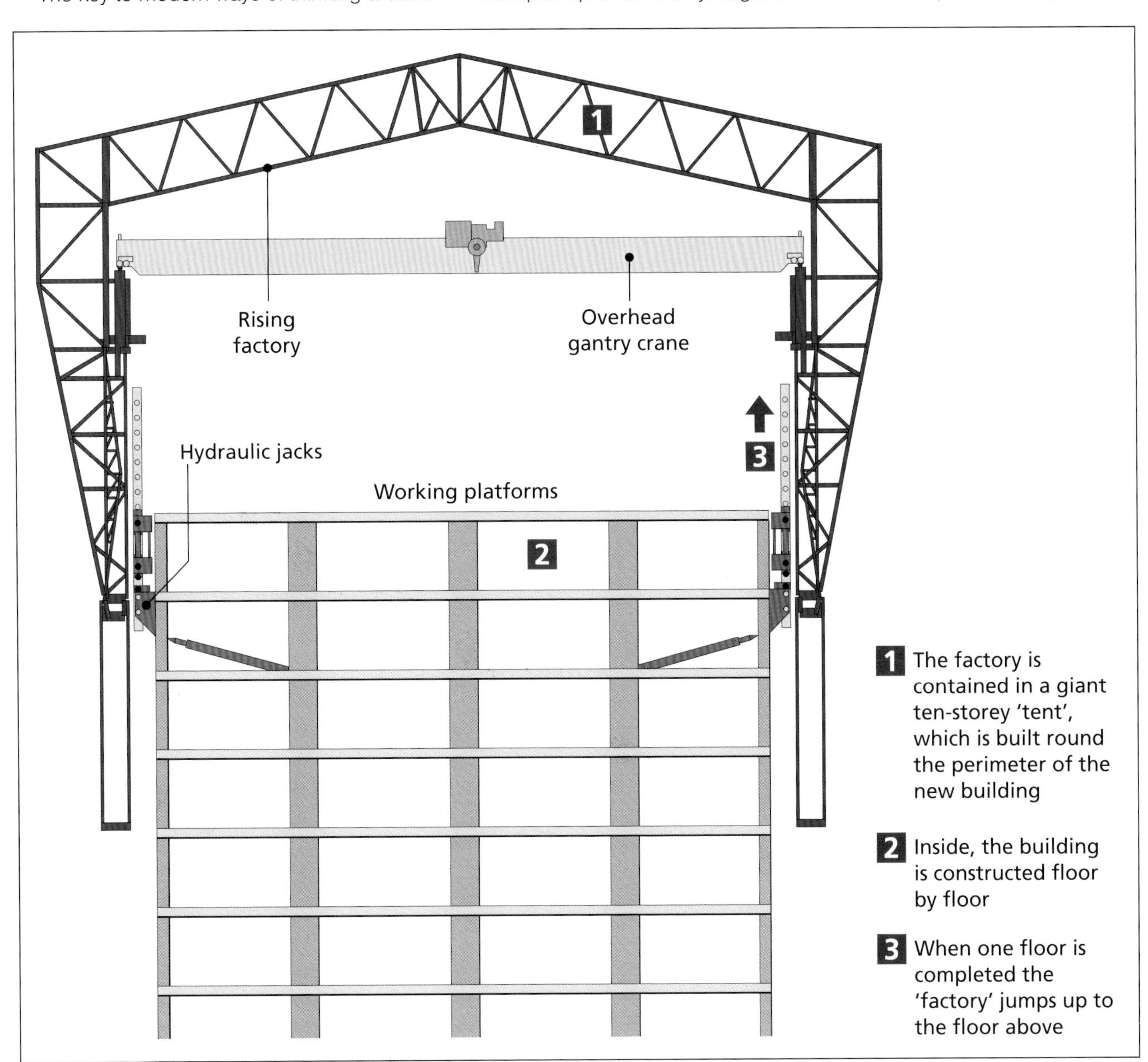

CAN WE 3D PRINT A SKYSCRAPER?

A 3D-printed skyscraper sounds like a fantasy. Perhaps one that conjures up a mental image of pressing a few buttons on a printer-like apparatus before a fully formed tower appears from a small output device in a matter of moments.

Well, that vision will likely remain a pipedream for some time to come. But 3D printing, or to give it its more accurate title 'additive manufacturing', is already making inroads in the world of construction, if not quite in skyscraper building just yet.

The technique refers to computer-controlled layering of materials to create 3D shapes. This technology began to emerge in the 1980s but struggled to make the associated effort and expense worthwhile. It has only become widely adopted since the start of this century.

The commercial benefits are now certainly being felt in smaller-scale manufacturing of electrical and plastic parts.

Over in the built environment, 3D printing can produce construction components and now even entire buildings.

3D-printed low-rise building technology is already a reality. Robotically controlled rigs can form structural concrete walls to which steel reinforcement is added in rapid time.

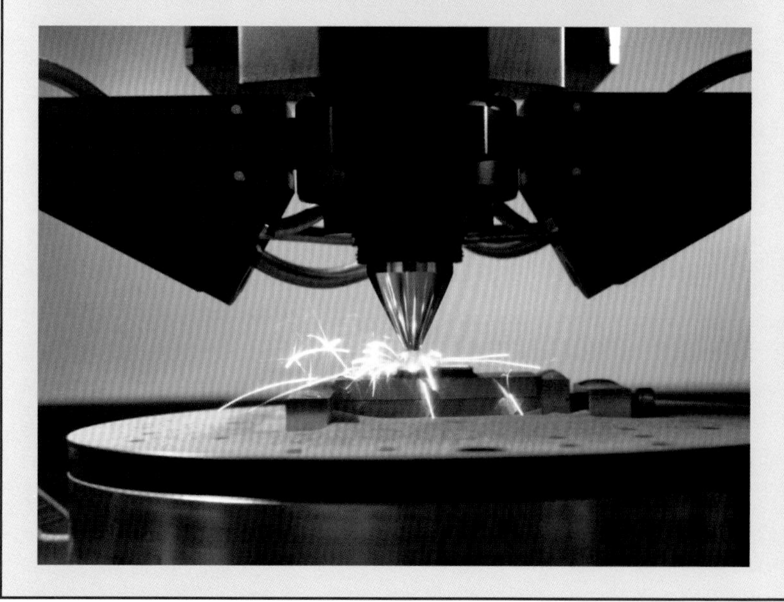

TOP Robotically controlled rigs are now able to form structural concrete walls to which steel reinforcement can be added. ... *(Shutterstock)*

CENTRE ... This technology has already resulted in 3D-printed low-rise buildings, and researchers are seeking ways to scale it up to potentially 3D-print a skyscraper in the future. *(Shutterstock)*

BOTTOM Refinements to 3D printing, or additive manufacturing, are still to be made, particularly in relation to how they handle and manipulate structural materials. *(Shutterstock)*

In response housing developers are now looking at adopting the technique on a broad scale to create replicable and affordable designs produced in a safe and speedy fashion.

There is also great potential for 3D printing to generate the many structural elements needed to enable complex geometric designs. Steel nodes that are designed to connect large structural frame elements together are a key example of where a boom could soon be seen using these methods. The advantages of faster and more accurate production allow for more bespoke items to be produced that would otherwise be difficult to mass produce using traditional manufacturing. Lower labour costs and less waste are additional benefits.

Whether a whole skyscraper will soon be 3D printed is yet to be seen. Taking risk out of projects by using manufacturing principles as well as removing construction workers from the often hostile or dangerous on-site environment is a tantalising prospect.

Refinements to additive manufacturing materials are still to be made, in particular surrounding how steel reinforcement is efficiently (and sufficiently) added to a 3D-printed concrete wall to make large-scale construction feasible.

And while one project hit the headlines in recent times for pledging to start building the first 3D-printed skyscraper soon (and no prizes for guessing that its location was to be in the Middle East), the blaze of public relations glory has since quietened down.

There are no firm plans in place to make the commercial and project risk realities worth the endeavour for skyscraper developers to embrace this technique in the immediate future. But the research and development continues and with big gains to be made it will not be too long before another attempt to be the first 3D-printed skyscraper is made.

working envelope, hugging the perimeter of the built structure below, are working platforms allowing personnel to carry out all the works to complete the building.

The main aim of the rising factory is to make building at height much more efficient, keeping working conditions enclosed by the tented structure and equipped with machinery and hoists to make it possible to build a whole storey in just hours rather than days or weeks. Each trade is organised to allow their activities to be carried out in specially designated areas on different levels within the factory.

Columns, internal walls, ventilation, drainage, mechanical and electrical services and even full bathroom and kitchen utility pods can be installed all on one level, before the entire factory is jacked up to begin work installing the floor above. As this continues, cladding work takes place on the floors below to make the lower storeys weathertight to enable first fix and fit-out to begin. Full fit-out and furnishing of the lower levels can even be completed while new storeys continue to be built above.

From the outside, as the factory jumps up, completed sections of the skyscraper are revealed below.

Towards the zero-carbon skyscraper

Until the 1980s few designers worried about the sustainability or carbon footprint of their tall-building projects. Constructing a conventional skyscraper consumed massive resources. Once complete, the occupants' comfort was largely dependent on giant, energy-hungry heating and ventilating units and banks of high-speed elevators. Wastewater and rainwater alike were simply dumped into the urban drainage system. But as energy prices climbed and worries about global warming began to enter the mainstream consciousness, new priorities began to emerge and ingenious new strategies were developed in response.

Attention first focused on the materials that made up the building. Could they be sourced closer to the project, minimising transport emissions? Could they include recycled or upcycled materials? Were there greener

OPPOSITE Once famous for the energy-saving thin-film real-gold coating to its curtain walls – costing more than $1 million at the time – Toronto's 1979 Royal Bank Plaza now uses the waters of Lake Ontario for cooling. *(Ajsdecpida)*

alternatives? Would going greener cost more money? And would more offsite manufacture and modular construction reduce the carbon footprint significantly? Answers are still coming forward, but more are needed.

Structural steel and other metals consume a lot of energy in their initial manufacture. They have to be mined, transported to the smelting works, crushed, heated in furnaces then rolled into I-beams, reinforcing bars and so on. Against this massive carbon footprint can be set the ease with which these metals can be recycled at the end of the building's life.

Concrete consists of a mixture of stones (coarse aggregate) and sand (fine aggregate) bound together by cement. Traditionally, this binder was Portland cement – and its manufacture contributes up to 7% of greenhouse gas emissions worldwide. The material is also relatively expensive. There have been a number of ways of reducing the Portland cement content developed over the last 50 years or so, originally with the objective of lowering the cost of the concrete. The durability of such modified concrete was usually improved as well.

Processed industrial by-products, such as slag from blast furnaces, ash from the chimney filters on coal-fired power stations and the very fine ash from the chimney filters on silicon metal furnaces, can replace significant percentages of the Portland cement content without loss of final strength. Now there are also advanced chemical admixtures that can improve strength and other properties while also reducing the Portland cement content.

Even with all these strategies, concrete has a problem with its carbon footprint. However, in locations where there are cement works nearby, as in the Middle East, while structural steel would have to be imported from another continent, concrete may well have the edge.

A more radical option is to go for a renewable structural material, such as timber or even bamboo (*see* Timber towers – a sustainable alternative, p.174, and Sky-high bamboo, p.176). All that is needed are open-minded developers prepared to balance initial cost against long-term running costs, and to care enough about global warming and the future of the planet.

There are many low-energy strategies available to create a comfortable environment for a skyscraper's occupants. High levels of insulation and the management of solar gain through adjustable blinds (*see* Keeping the weather out, p.52) is one example. Conserving the energy within the building is another, which can offer dramatic benefits.

Elevators can be fitted with regenerative motors, similar in principle to those coupled to the main engines in the current crop of F1 cars. As they move upwards, the electrical energy that lifts the cars is transformed into potential energy. As they descend, that potential energy is converted back to electric energy, which is fed into the building's energy grid.

Warm 'grey water' from bathrooms and kitchens can be passed through a heat exchanger to preheat incoming mains water. It can then be stored and used for toilet flushing.

Below the building the earth will be at

BELOW Heat pumps can concentrate heat from the ground or even aquifers. *(Anthea Carter)*

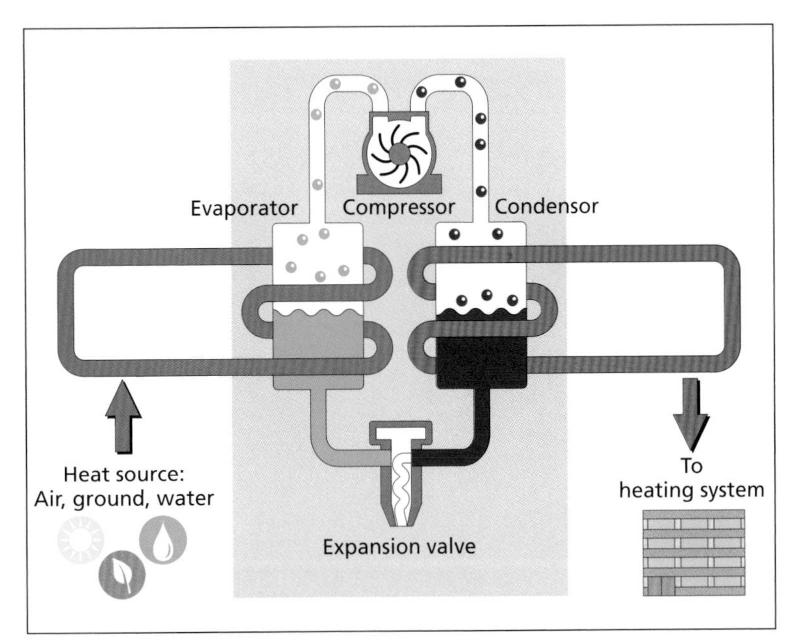

LEFT London's Heron Tower has a façade-mounted solar photovoltaic array. *(Eluveitie)*

a fairly constant temperature all the year round. In summer it will be cooler than air temperature, in winter it will be warmer. These temperature differences can be exploited by various means to minimise the need for extra heating and cooling throughout the year. Heat pumps – using the same basic technology as refrigerators – can extract and concentrate heat from the ground in winter and reverse the process in summer, storing the heat for winter use. Some conveniently located buildings can take advantage of nearby deep lakes or even subterranean aquifers.

Solar photovoltaic (PV) cells have been successfully installed on skyscraper façades. There are even transparent PV cells available, although the current generation are not capable of providing a really significant alternative to a connection to national energy grids.

A few recent skyscrapers have been designed with integrated wind turbines. These have not yet demonstrated the ability to contribute serious percentages of the building's energy needs, however. The basic problem is that to generate significant amounts of electricity a wind turbine needs to be at least 30m (98ft) in diameter on a 40m (131ft)-high tower, located in a smooth airflow.

Bolting such an assembly to the top of a skyscraper would feed high loads and vibration into the structure. And as was discovered when airships attempted to moor up to the mast that topped the Empire State Building (*see* Empire State Building, 1931, p.126) there is usually severe air turbulence above a tall building. This turbulence also militates against the concept of installing a cluster of smaller wind turbines on the roof. Even in a smooth airflow region, small turbines in close proximity create enough turbulence to seriously reduce their efficiency.

A well-designed skyscraper that takes advantage of energy-saving and conservation techniques will frequently have surplus heat and electrical energy to deal with. Surplus electricity can just be fed back into the national grid – an attractive option when generous

RIGHT **Integrated wind turbines were expected to generate up to 8% of London's Strata SE1 electricity needs. Results since it opened in 2010 are understood to be disappointing.** *(Colin)*

feed-in tariffs are available. Surplus heat can be stored for use in winter. Using the ground below the building is one option as already mentioned; another is to install dedicated energy stores.

These can use advanced phase change materials – or be as simple as an insulated tank of water. Some stores are intended just to hold heat overnight, others for an entire summer. If there is no financial benefit in transferring surplus electricity to the national grid, it could instead be used to add extra heat to the energy store.

It is unlikely that even the most advanced skyscrapers will ever be totally independent of national electricity and potable water grids. Yet this dependency and the associated carbon footprint can be minimised by designs that put sustainability high on the priority list. Cynical 'greenwashing' must be avoided. Strapping a token wind turbine or a small yet surprisingly visible solar PV array on to an otherwise conventional tower is not the way forward.

A proper evaluation of a skyscraper's sustainability must also look at potential maintenance costs, and the cost of demolition or refurbishment when the building has reached the end of its viable life. Ideally, there should be as much recycling as possible. Even high-strength concrete can be crushed and processed into coarse aggregate, while structural steel sections can be reused on new projects, as can engineered timber, high-performance glass and a wide range of fixtures and fittings.

This reuse is a feature of the 'circular economy', which prioritises reuse over simple recycling. Anything that helps reduce greenhouse gas emissions should be embraced with enthusiasm, given the increasingly gloomy predictions of the effects of global warming. A skyscraper whose overall carbon footprint, calculated over its lifetime, is as close to zero as possible, should be the objective of every responsible designer.

Since the turn of the 21st century there have been a number of medium-rise buildings constructed from timber, although none has yet topped the 100m (328ft) mark to become the world's first timber skyscraper. There have been taller timber structures: a 1930s transmitter tower in southern Germany reached 190m (623ft), while Lincoln Cathedral's record-breaking wooden spire was 160m (525ft) tall (*see* World's tallest pre-20th-century structures, p.16).

Both of these long-gone record-holders were constructed from traditional sawn hardwood. Sawn timbers have a number of drawbacks, not least the difficulty of obtaining large enough timbers for major projects. The inherent variability of sawn timber also makes it very hard to carry out structural analysis to modern design codes. As a result, construction in sawn timber has tended to be more of a craft than an engineering discipline.

All this changed with the development of durable, high-strength synthetic resin adhesives in the second half of the 20th century. This made it possible to glue thin timber veneers or small sections together to create a wide range of 'engineered' timbers, the first of which was the ubiquitous plywood. Then came glued laminated timber, better known as glulam, where precisely cut timber sections are aligned with their grains parallel and bonded together under pressure to form large structural members.

Maximum size is limited only by the logistics of transport. Architects in particular have taken advantage of the ease with which glulam can be shaped into a wide variety of structural forms, such as arches and cantilevers. Glulam columns have been at the heart of a number of tall timber building proposals – but it was the advent of cross-laminated timber (CLT) panels that was the real game-changer.

Softwood planks up to 50mm (1.97in) thick and 240mm (9.45in) wide are glued and stacked up to seven layers thick, with the grain in each layer running at right angles to the next. Again, bonding takes place under pressure, and maximum size is limited by transport logistics.

Large CLT panels are usually delivered to site in a fully prefabricated state having been accurately cut to size and shape on computer-controlled machines, with all doors, windows, service penetrations and joints preformed. Assembly is rapid, quiet and largely dust-free, which has made CLT construction a popular choice for medium-rise apartment blocks in sensitive urban areas.

Seven-ply CLT can be 400mm (15.8in) thick and is able to take serious loads. At first, designers were cautious, and timber towers rarely rose higher than 8 or 9 storeys, with the tallest just 14 storeys. Some of these are better described as hybrid structures, as they retained the traditional reinforced concrete central service core. However, as sustainability pressures intensified, some designers began to seriously consider engineered timber as an alternative to conventional materials for much taller towers – true skyscrapers.

One contender is the 300m (984ft)-tall, 80-storey Oakwood Tower proposed for a site in central London. Developed by the team of structural engineer Smith and Wallwork and architect PLP Architecture, with researchers from Cambridge University's Department of Architecture working alongside, the tower would be the second tallest in London after The Shard and would be constructed almost entirely from glulam and CLT. The mixed-use building would include more than 1,000 residential units.

Massive glulam sections are at the heart of the design. With columns measuring 2.5m by 2.5m (8.2ft by 8.2ft), and truss components up to 35m (115ft) long, weighing in at 100t (110 US tons), engineered timber technology would be pushed to new limits. There has been little if any research into the performance of such enormous glulam sections, and the logistical challenges would be daunting.

What the designers describe as 'mega-carpentry' could be used to form the largest glulam elements on site from smaller, easier-to-transport sections. Steel nodes would connect the completed sections together.

A buttressed mega-truss (*see* Standing tall – the structural frame, p.82) was the chosen structural system. One of the challenges for all timber skyscrapers will be how to resist overturning forces from wind loads. In conventional skyscrapers the sheer mass of concrete and steel in the building and its foundations is easily able to cope with such forces – but a timber skyscraper would be much, much lighter.

However, ensuring that the lateral load-resisting system is concentrated at the building's perimeter solves the problem. It also has a secondary advantage. The service core, freed from a lateral load-resistance function, can be located anywhere convenient within the building.

Another advantage would be the locking-in of 50,000t of CO_2 in the structural timber frame. This would use 65,000m^3 of softwood from certified forests, and the proposers point out that softwood forests, unlike tropical rainforests, are currently increasing in size.

There may be understandable concerns regarding the fire resistance of all that softwood. Apart from the fact that any new skyscraper would be fitted with automatic sprinkler systems, engineered timber enthusiasts point to the extensive research into the fire performance of timber buildings, including full-scale tests. These show that the massive timber structure of the Oakwood Tower should actually perform *better* in a fire situation than a steel equivalent. A layer of insulating char forms on the timber surface as flames attack it, leaving the main structural section still more than capable of bearing the loads.

Exposed timber in a building is highly popular with its occupants – which is why barn conversions with oak beams and stripped and polished floorboards are seen as highly desirable properties. CLT floors are proposed for the Oakwood Tower, but there is still research needed into how such long-span CLT floors would react to footfall. Or to put it another way will footsteps from the floor above be annoyingly audible to those living below?

Recent research has also shown that timber buildings can have beneficial effects on their occupants' health and well-being. Given all the additional advantages of timber, not least its sustainability, it is hardly surprising that many believe it will be the signature structural material of the 21st century.

RIGHT **Despite its name, the Oakwood Tower would use 65,000m^3 of sustainable softwood in its engineered timber structure.** *(With thanks to PLP Architecture, Smith and Wallwork and Cambridge University)*

SKY-HIGH BAMBOO?

Far Eastern construction sites have been viewed as primitive to Western eyes. Instead of the modular steel scaffolding universal in the West, a forest of triangulated bamboo canes envelopes many tall structures, with the individual canes simply lashed together with ropes and cords. Such scaffolding might appear primitive at first glance, but in practice problems are rare.

Bamboo as a structural material has a surprising number of advantages. It can resist twisting loads (torsion) better than most structural steels, so it performs better in earthquakes. Its tensile strength matches that of steel, too. Weight for weight, the lightness of bamboo gives it superior structural properties. Its only real weakness is the difficulty of connecting bamboo elements together: it has a tendency to split if holes are drilled in it.

Best of all, however, is bamboo's sustainability. Basically a giant grass, it is the fastest-growing plant on the planet. In just three years some species can reach a height of more than 30m (100ft) and a diameter of 200mm (8in) or more. Over the centuries the Chinese have used bamboo to make everything from paper to musical instruments to the very first gunpowder weapons. In the West, split bamboo fishing rods were the best available until the advent of carbon fibre. Bamboo is a renewable resource with a very low-carbon footprint – no wonder designers are beginning to look seriously at its potential.

Sustainability became an increasingly important priority for skyscraper designers as the 21st century dawned (*see* Towards the zero-carbon skyscraper, p.169). Many approaches were developed, but among the most adventurous designers there were serious proposals for building tall with bamboo. Sustainability was more than just substituting bamboo for steel or concrete. A bamboo tower would be significantly lighter than conventional designs, meaning smaller, cheaper foundations and an even lower carbon footprint.

There are many desirable locations for skyscrapers where the ground conditions are such that very deep-piled foundations would be needed to support a typical steel-

RIGHT Just an architects' vision at the moment – but the age of the bamboo skyscraper may yet dawn. *(Courtesy of CRG Architects)*

or concrete-framed structure. This could blight any high-rise development proposals. A lightweight bamboo tower, by contrast, could need only a simple concrete raft – quick and cheap to install.

Traditional jointing methods used for permanent constructions are remarkably sophisticated, even though they almost always depend on some form of lashing to hold the joints together. Skyscraper designers are unlikely to be willing to go down this route. To quote Beijing-based CRG Architects: '200m [656ft]-high buildings cannot be tied with ropes!'

CRG opted for stainless steel nodes in the design of the extraordinary Architect's Village, a shortlisted entrant in the 'Future Projects' category of the 2015 World Architecture Festival Awards. This was a forest of circular towers up to 200m (656ft) tall, based on an external bamboo diagrid – a concept inspired by natural bamboo forests.

Central service cores are also based on a bamboo diagrid. The bamboos selected would be about 30m (98ft) long, and the structures would be weatherproofed by an external translucent ethylene tetrafluoroethylene (ETFE) membrane.

Public acceptance of such a revolutionary concept may be helped by reassurances on the fire resistance and durability of a bamboo structure. There are many techniques, both ancient and modern, that improve bamboo's properties. These range from six months' immersion in seawater through simple borax solutions to thin film intumescent coatings (*see* Dealing with disasters, p.110). These protect the bamboo against its worst enemies – fungus and insects – and increase its fire performance to modern regulatory standards.

Bamboo structures adequately protected against the worst of the weather can last a long time if the bamboo has been prepared properly. There are centuries-old bamboo structures still standing in Asia. Although the day of the 200m (656ft)-tall bamboo skyscraper has yet to dawn, there is a glimmer of light on the horizon.

ABOVE The Twin Towers' iconic status made them a terrorist target. *(Jeffmock)*

Lessons from 9/11

What happened in New York on 11 September 2001 shocked the world. More Americans died that day than in the notorious Japanese attack on Pearl Harbor in 1941. It was unprecedented, unthinkable. Yet before the clouds of toxic dust created by the collapses had settled, the question was being asked. How could two modern skyscrapers be brought down so easily?

Speculation was rife and conspiracy theories were prominent. The Twin Towers were too light, too fragile, said some. It was all a sinister plot by US President George W. Bush and the towers were actually destroyed by hidden demolition explosives, said others. The country's National Institute of Standards and Technology (NIST) was commissioned to carry out a detailed investigation into every aspect of the catastrophe, although it was obvious from the start that some would never accept any conclusion it came to.

When they opened in the early 1970s the Twin Towers of the World Trade Center in Lower Manhattan were seen as a dazzling feat of design and construction. Using the revolutionary principles pioneered by world-famous structural engineer Fazlur Rahman Khan, the towers were 40% lighter than conventional designs and extensive prefabrication meant that erection proceeded at a dizzying pace.

Construction of the North Tower began in August 1968, and the first tenants moved in less than 30 months later. The South Tower started later, completing in January 1971.

Khan's development of the 'tube in tube' and 'outrigger and belt truss' concepts were the inspiration for the Twin Towers' design. The perimeter of each tower would be defined by 59 closely spaced square section columns on each face. These measured 360mm (14in) square and were fabricated from welded steel plates. Between them, narrow windows just 450mm (17.7in) wide made up only 30% of the façades.

Unlike many tall towers, there were no reinforced concrete central cores housing the buildings' elevators and services. Instead, there were central grids of full-height box-section columns, within which were the elevator shafts and three emergency stairwells.

Concrete cores have the benefit of offering high levels of fire protection to the emergency stairs, and to vital services such as sprinkler supply pipework. For the Twin Towers, however, the designers specified nothing more substantial than fire-resistant gypsum wallboard to clad the central core columns.

Floor design combined lightness with speed of erection. Prefabricated lightweight steel

BELOW These lightweight floor trusses are seen as the most likely to have failed first in the fires. *(Anthea Carter)*

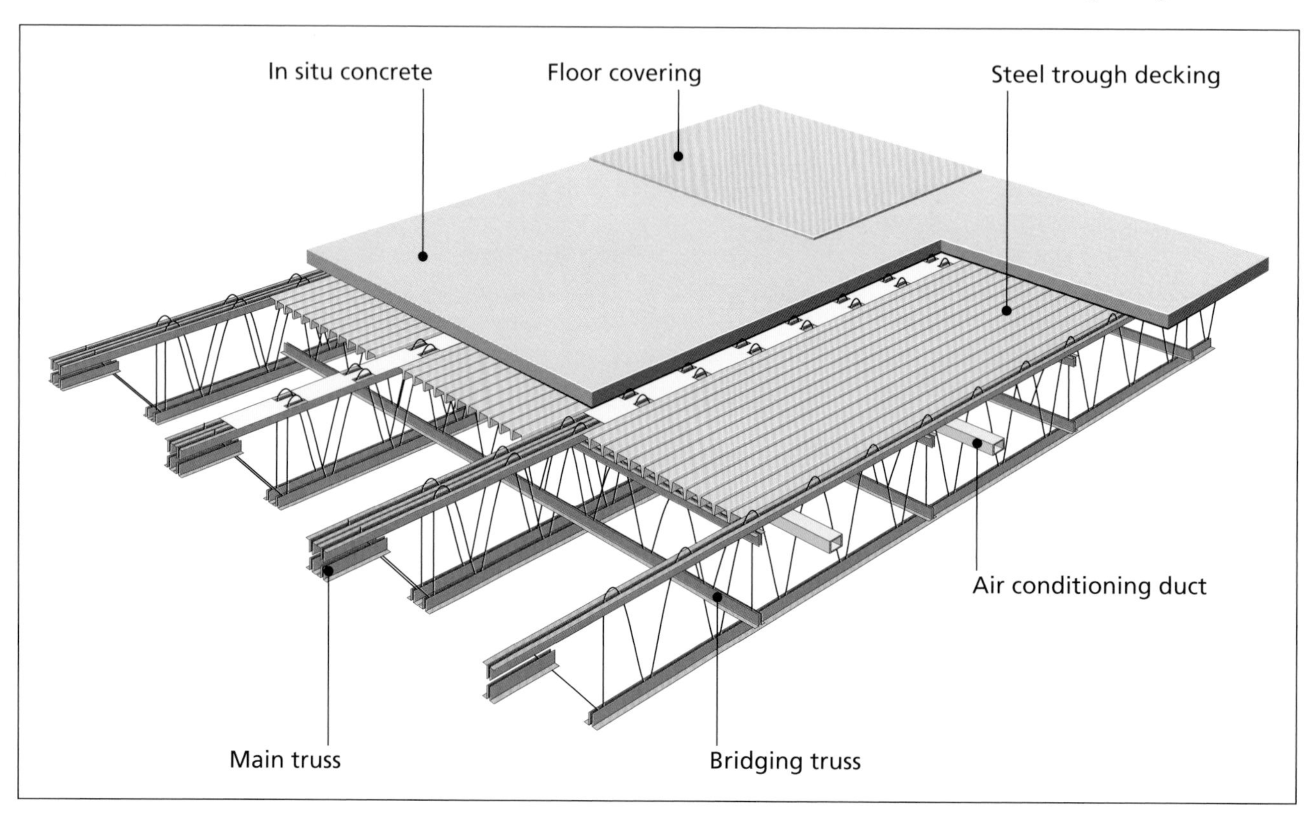

trusses topped with profiled galvanised steel moulds spanned 18m (59ft) between the core and exterior columns. Viscoelastic dampers – basically pistons in metal housings filled with a highly viscous fluid – were installed at the connections between the floor trusses and the exterior columns. The floor moulds were filled with lightweight concrete – that is, concrete in which the heavy stone aggregate in normal concrete has been replaced by low-density expanded clay pellets.

Core and perimeter columns were connected between the 107th and 110th floors by ten steel belt trusses. All internal steelwork had a spray-applied cementitious fireproofing system designed to protect the structure for at least 120 minutes. There is still debate over whether the fireproofing was thick enough or evenly applied.

Originally the towers had no automatic sprinkler systems, as the relevant building codes of the time did not require them. However, an arson attack in 1975 caused $1 million of damage, so in 1981 the decision was taken to retrofit a sprinkler system over a period of three to five years.

To minimise the number of elevators needed, the designers introduced 'sky lobbies' (*see* People movers, p.101), with two elevators in most shafts. This was a decision that saved lives during 9/11.

A tube is a very resilient structural form. When the Boeing 767s punched holes through the tube wall, stresses were redistributed to the many remaining columns. The wrecked fuselages also penetrated as far as the cores, severing or distorting an unknown number of core columns. The towers absorbed the impacts, and remained standing.

Principal structural engineer Lesley Robertson said later that the design team at Worthington, Skilling, Helle & Jackson had analysed their design to see if it could cope with a major aircraft impact. For the analysis they assumed a Boeing 707 – which weighed a little less than a 767 – would be lost in fog while making an airport approach and would hit a tower at around 240kph (150mph). The answer was reassuring and was tragically confirmed more than three decades later.

Robertson also added that if the hijacked aircraft had been 300t (330 US ton) Boeing 747s rather than the 150t (165 US ton) 767s, 'the tops of the towers would have been knocked off'.

What the design team never considered would be the effects of thousands of litres of flaming jet fuel cascading down through the building.

So flammable is jet fuel that it was all consumed in a few minutes. Those few minutes sealed the fate of the towers. The fuel acted as a gigantic firelighter, igniting the office contents on several floors at the same time. These burned for more than one hour.

It is almost certain that the automatic sprinkler systems were knocked out by the initial impacts. Even if they had survived, it is doubtful they could have coped with simultaneous widespread fires on several floors. It is also likely that the enormous shock of the impacts jarred the brittle cementitious fireproofing off large sections of the columns and the floor trusses.

What is known as the fire load for office floors is high, thanks to the presence of lots of paper, plastics and furniture. Temperatures at ceiling level above a burning office floor rapidly exceed 1,000°C (1,832°F). Steel, as we have seen, softens at around 450°C (842°F). By the time its temperature tops 650°C (1,202°F) it will have lost at least half its strength. In the Twin Towers it would be the thin steel sections making up the floor trusses that would have heated up the fastest, especially if the fireproofing had been lost.

Simultaneously there must have been some complex interactions between the core columns above the impact zones, the belt trusses and the perimeter columns. These could have transferred extra loads into the perimeter columns, the magnitude of which would be related to how many core columns were no longer able to carry loads.

As the fires continued to blaze, floor trusses began to sag. Eventually the perimeter columns began to soften. The weight of the sagging floors pulled the softened columns inwards, buckling them and triggering progressive collapse.

Demolition contractors sometimes deliberately trigger progressive collapses when they have to demolish a low- to medium-

rise building. They do this by exposing all the columns on a lower floor, then cutting through a significant number before wrapping the remaining columns with high explosive.

All these explosive charges are then linked together by a spider's web of cables. All the charges must detonate at the same millisecond if the building is to collapse straight down and not topple sideways, perhaps into adjoining buildings.

It is hard to imagine how such demolition preparations could have been carried out in the Twin Towers without anyone noticing, or how the impact of the planes could be arranged to be on the same floors as the hidden demolition charges.

Nevertheless, the striking similarity between the 9/11 collapses and the videos of deliberately induced progressive collapses convinced many people that the official reports were a calculated deception, a cover-up. Nearly two decades on, controversy continues.

Initial attention focused on how the hijackers managed to board the airliners armed with sharp utility knives – the notorious box cutters. At a peak in the 1960s an average of one airliner was hijacked in the USA *every six days.* This was down to almost non-existent airport

OPPOSITE Later analysis showed that the massive dust clouds generated by the collapsing Twin Towers were a major health hazard. *(Wally Gobetz)*

BELOW Destruction on this scale triggered many conspiracy theories. *(NOAA)*

security at the time. Luckily, none of those hijackers were on suicide missions, but the situation became so extreme that eventually strict new security procedures were introduced and hijackings dwindled to almost zero.

Even stricter security procedures were hastily introduced after 9/11 and remain in force to this day. Airlines installed armoured locking doors between the cockpit and the passenger cabin, with a video link to show pilots who might be seeking entry. However, as the crash of Germanwings Flight 9525 in March 2015 tragically showed, a suicidal pilot can lock the door against the rest of the crew and deliberately fly the plane into a mountainside – or a skyscraper.

Structural engineers were well aware that the inevitable demands for stronger, safer skyscrapers would be virtually impossible to satisfy. If 767s could bring down the Twin Towers, what havoc might a 500t (551 US ton) Airbus 380 with a suicidal pilot wreak?

This was not something those who lived and worked in skyscrapers wanted to hear. There was widespread anxiety, even organised tenant protests. When it was revealed that the Twin Towers had only three sets of escape stairs each instead of the six that were the norm in more modern skyscrapers, there was outrage. The spine-chilling stories told by the survivors, of congestion and confusion on the stairs, of firefighters attempting to climb up burdened with heavy equipment having to force their way through downrushing waves of desperate evacuees, of the Herculean task of moving wheelchair users down 60 or more flights of stairs, fuelled the anger.

There was an intense focus on means of escape in the event of a major catastrophe. The conventional clustering of escape stairs into a central core was heavily criticised. Another airliner impact, a suicide bomber, a major fire, even an earthquake – these could knock out all the escape stairs at once, leaving occupants once again trapped on the upper floors.

One suggestion was to have at least two sets of widely separated escape stairs, perhaps at diagonal corners. These, unfortunately, would not be an optimum structural solution. One thing was certain, however. No future skyscraper would rely on flimsy gypsum wallboard to protect its emergency stairs.

What 9/11 survivors reported made it very clear that the emergency stairs there were too narrow to cope with the sheer volume of occupants attempting to escape down them. Those who put their own lives at risk to save wheelchair users caused significant congestion as they slowly descended the stairs with their burdens. Those firefighters trying to force their way upwards also caused delays, all because of the narrowness of the stairs.

In the South Tower a few survivors from above the impact discovered that the emergency stairway below the sky lobby at the 78th floor was still largely intact and accessible. The elevator mechanism in the shaft had offered significant protection to the lobby area – unfortunately, though, most of the occupants on the upper floors failed to discover this.

There was another design flaw. Those attempting to enter the stairs from lower floors were confronted head on by people descending from above. Many had difficulty joining the crowd heading for the next flight down (*see* diagram opposite). Simply moving the entrance door sideways so it aligned with the downward flight would make entry far less restricted.

Building codes and regulations worldwide were rapidly amended after 9/11. The minimum width for escape stairs was generally increased by 25%. Dedicated firefighter access stairs and elevators became the norm. There was still one major problem needing resolution, however.

John Abruzzo, a severely disabled accountant working on the 69th floor of the North Tower had an amazing story to tell. Reliant on an electric wheelchair, he was there in 1993 when the truck bomb detonated in the basement (*see* Dealing with disasters, p.110). It took a long six hours after the explosion until he was finally evacuated. As a result of his experience, and that of others, special lightweight 'evacuation chairs' were made available.

On 11 September Abruzzo was evacuated in one such chair, carried by three or four people. It took 90 minutes to get him down the 69 floors, and he and his rescuers escaped just 10 minutes before the North Tower collapsed.

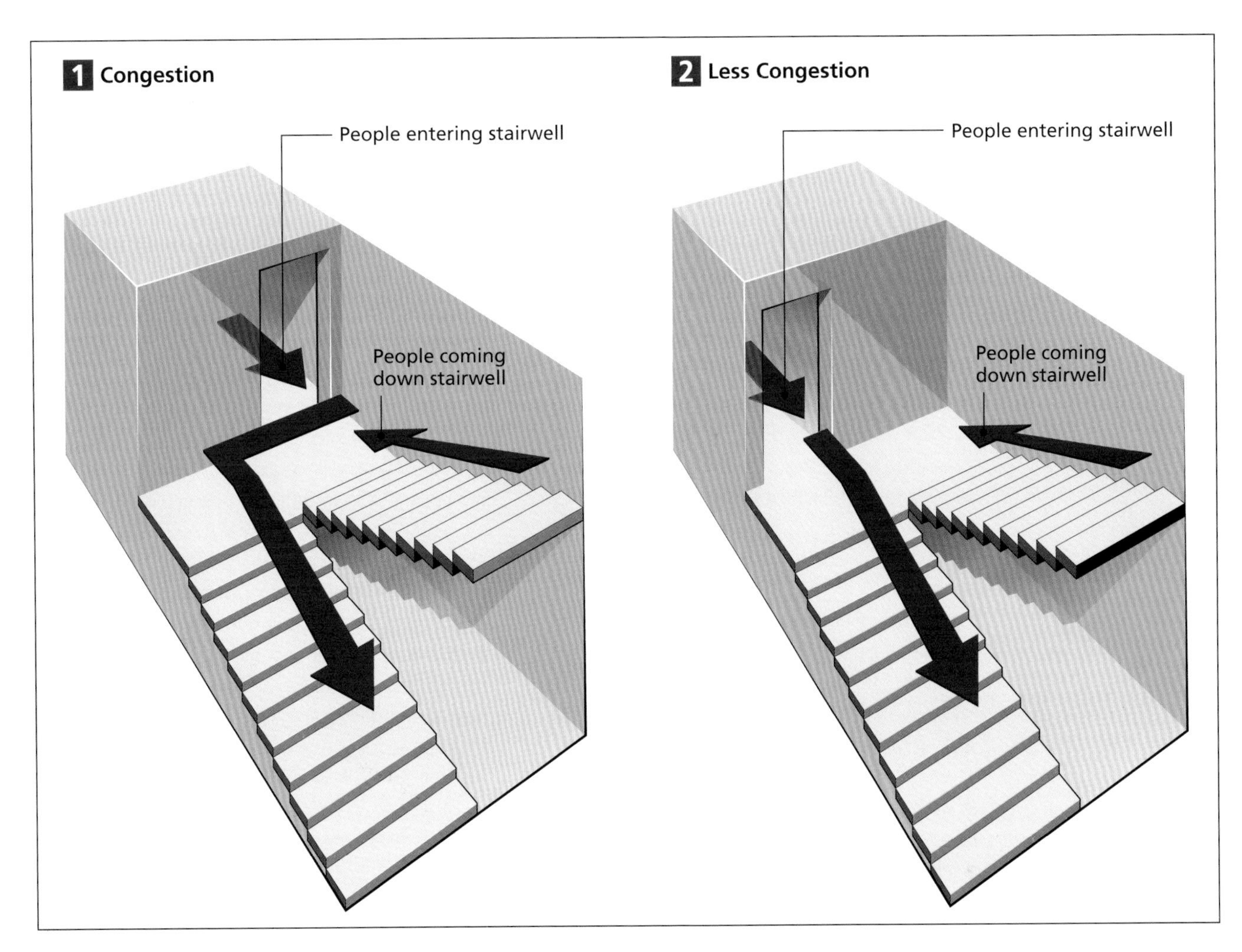

ABOVE A simple revision to escape-stair entrances would significantly reduce the risk of dangerous congestion during an evacuation. *(Anthea Carter)*

A heroic escape, no doubt. But, asked some, how many people were trapped as a result of escaping wheelchair users slowing their descent?

Attention turned on the role of the elevators. From the earliest days of the skyscraper, occupants had been warned repeatedly not to use them during a fire. As buildings grew ever taller, this policy began to look ever more unrealistic.

Initially the focus was on using elevators – perhaps the dedicated firefighter elevators – to evacuate wheelchair users only. Then it was pointed out that there would be many occupants who, through age or chronic ill health, would find descending up to 100 floors on foot almost impossible, certainly requiring several rest periods with their potential for increasing congestion.

Very tall buildings, such as the Burj Khalifa, have adopted a number of emergency options. Dedicated firefighter elevators are standard, but many very tall buildings now include a proportion of fire-resistant elevators entered through protected lobbies that can be used for general evacuation. Comprehensive internal communications and regular safety drills for occupants are also universal.

Some still question whether these precautions would work in an actual emergency. People are unpredictable. Research indicates that willingness to use them in a real emergency cannot be taken for granted. Occupants on higher floors are probably more likely to turn to the elevators rather than face the long, hazardous descent via the emergency stairs. A lot seems to depend on how reliable and how fast the elevators are perceived to be in that particular building.

Postscript

On 14 June 2017 fire ravaged Grenfell Tower, a 67m (220ft)-tall local authority-owned residential tower block in west London; 72 people died, and a further 70-plus were injured. It was the worst residential fire disaster in the UK since the Second World War.

This was a unique event. No other tower block fire anywhere in the world spread so rapidly throughout the building. No other residential tall building ever had all four façades on fire simultaneously. And in no other tall building fire has there been such controversy over the causes, the response of the fire service and the role of government regulations.

At the time of writing (January 2020) the final verdict on the tragedy is still awaited. Much information has been released, however; enough to reach some preliminary conclusions. Already there has been urgent work carried out on many similar tower blocks in the hope that a similar disaster can be avoided. What the final impact on the design of high-rise residential buildings might be is still not clear, as are the long-term implications for residential and mixed-use skyscrapers.

What was a matter of record was that the building had undergone a major refurbishment both internally and externally between 2015 and 2016. To reduce heating costs, a 150mm (5.9in)-thick layer of polyisocyanurate (PIR) foam was bonded to the original reinforced concrete cladding. This was then protected by a rain shield of aluminium composite material (ACM) – basically two thin skins of aluminium either side of a 3mm (0.1in)-thick polyethylene core.

This particular type of ACM was not recommended for use on tall buildings by its manufacturers, who also produced more fire-resistant (and more expensive) ACM options.

Tower block fires are thankfully rare; fatalities in such fires generally are low. Most follow a similar pattern. A fire starts within an individual flat or apartment. Each flat is in effect a fire-resistant box that can contain the flames for at least 90 minutes. This 'compartmentation' gives firefighters enough time to arrive on site, extinguish the fire and prevent any internal spread to other apartments.

There is one weak spot, however. As temperatures rocket inside the burning apartment, the window will eventually shatter and allow flames to burst outwards and upwards. An aerodynamic phenomenon known as the Coanda effect keeps the flames close to the façade, which means they soon are raging outside the window of the apartment above.

Radiation from the flames passes through the window and ignites the contents of the second apartment. Eventually flames from this source will be burning outside the window of the next flat above – and so the fire progresses.

This is a relatively slow and essentially vertical progression, which rarely proceeds more than a few floors upwards before firefighters have the blaze under control. This scenario, however, assumes that the internal compartmentation is intact and functioning as intended.

At Grenfell the speed at which the flames spread from apartment to apartment astounded and shocked even fire safety experts. Attention focused first on the ACM, where the thin aluminium simply melted, and the core burned. Some of the PIR insulation also burned – but there were large areas where it simply charred and did not ignite. There were those who wondered if the 3mm (0.1in) of polyethylene in the ACM had enough thermal energy to ignite flats from the outside, and whether the main cause of the rapid spread was down to a massive failure of internal compartmentation.

Independent full-scale fire tests on the Grenfell cladding system also showed little tendency for lateral spread of fire. Then another possible contributory factor came to light. Replacement windows had been installed, windows that were said to be smaller than the originals, and thus needed larger surrounds to fill the gaps.

These surrounds could have had little fire resistance, allowing flames to spread in both directions much too easily.

In late 2016 residents had warned that the building's maintenance was inadequate and that health and safety laws were being ignored. They also singled out exposed gas pipes as a particular area of concern.

Checks on similar tower blocks in the aftermath of the tragedy revealed many with similar types of cladding. Some serious shortcomings in internal compartmentation, were also uncovered. Recladding has already begun in advance of the final report.

There are those who believe that retrofitting automatic sprinklers to vulnerable buildings would be less disruptive than recladding in many cases. What is almost certain is that if Grenfell Tower had been equipped with sprinklers, nobody would have ever heard of it because the original fire would have been extinguished at source. And 72 people would not have died.

ABOVE The worst residential fire disaster in the UK since the Second World War. *(Public domain)*

Index